Future Remains

The University of Chicago Press, Chicago 60637
The University of Chicago Press, Ltd., London
© 2018 by The University of Chicago
Published 2018
Printed in the United States of America

26 25 24 23 22 21 20 19 18 17   1 2 3 4 5

ISBN-13: 978-0-226-50865-8 (cloth)
ISBN-13: 978-0-226-50879-5 (paper)
ISBN-13: 978-0-226-50882-5 (e-book)
DOI: 10.7208/chicago/9780226508825.001.0001

Library of Congress Cataloging-in-Publication Data

Names: Mitman, Gregg, editor. | Armiero, Marco, 1966– editor. | Emmett, Robert S., 1979–
    editor.
Title: Future remains : a cabinet of curiosities for the Anthropocene / edited by Gregg Mit-
    man, Marco Armiero, and Robert S. Emmett.
Description: Chicago ; London : The University of Chicago Press, 2017. | Includes index.
Identifiers: LCCN 2017009032 | ISBN 9780226508658 (cloth : alk. paper) | ISBN
    9780226508795 (pbk. : alk. paper) | ISBN 9780226508825 (e-book)
Subjects: LCSH: Nature—Effect of human beings on. | Nature and civilization. | Human
    ecology.
Classification: LCC GF75 .F88 2017 | DDC 304.2—dc23
LC record available at https://lccn.loc.gov/2017009032

♾ This paper meets the requirements of ANSI/NISO Z39.48-1992 (Permanence of Paper).

and Robert S. Emmett

Future

Remains

A Cabinet of Curiosities

Contents

Acknowledgments

Future Remains, in its multistage format, including a slam, Cabinet of Curiosi- vii
ties exhibition, and writing workshop, has been a particularly collective effort.
The Nelson Institute's Center for Culture, History and Environment (CHE) at the
University of Wisconsin–Madison, the Rachel Carson Center for Environment
and Society at LMU Munich, the Deutsches Museum, and the Environmental
Humanities Laboratory at the KTH Royal Institute of Technology, Stockholm,
supported this project from its inception at a meeting in Munich in August 2013
into the present book form. In addition to the fantastic group of contributors represented in these pages, we also want to thank all the participants and audience
members in Madison and Munich who contributed ideas and creative energies
through the wildly open format of the Anthropocene Slam in 2014 and "Welcome to the Anthropocene: The Earth in Our Hands" exhibit at the Deutsches
Museum. We are grateful for the work of the original Environmental Futures
planning team, including Samer Alatout, Bill Cronon, Wilko Graf von Harden-
berg, Richard Keller, Christof Mauch, Anne McClintock, Sabine Moedersheim,
Rob Nixon, Lynn Nyhart, Marc Silberman, Sverker Sörlin, Helmuth Trischler,
and Nina Wormbs, as well as the funding support made possible through a generous grant from the German Academic Exchange Service (DAAD) to the Center
for German and European Studies at the UW–Madison.

A special thanks go to Garrett Dash Nelson for his heroic organizational
efforts, to other members of the Anthropocene Slam selection committee, includ-

ing Heather Swan, and to our hosts in Madison in the fall of 2014, including Bill Cronon, and the CHE faculty, graduate students, staff, and community members. We are grateful to the Anonymous Fund, the Center for the Humanities, the Nelson Institute's Center for Climatic Research, the Nelson Institute's Center for Sustainability and the Global Environment, and the Nelson Institute for Environmental Studies at UW–Madison for their help in supporting Elizabeth Kolbert's keynote lecture as part of the Anthropocene Slam. The Madison event would not have been possible without the support of Nelson Institute director Paul Robbins, and the efforts and talents of Nelson Institute staff including Danielle Lamberson Philipp, Andrew Ortman, Steve Pomplun, Hope Simon, and Ann Swenson.

At the Rachel Carson Center (RCC), Christof Mauch and Helmuth Trischler provided input and support at critical moments, while Daniela Menge, Kim Coulter, Iris Trautmann, Carmen Dines, Marie Heinz, and Annka Liepold helped translate the conceptual cabinet into the museum and virtual exhibits (on www .environmentandsociety.org). The RCC also supported a pivotal writing workshop held in July 2015 in Munich. Without generous research and travel support from the DAAD and Federal Ministry for Research and Education (BMBF), the international scope of this collaborative project would have been impossible. Special thanks to Lynn Keller and the CHE Steering Committee along with the KTH Environmental Humanities Laboratory for making possible the production and printing of the color plates in the book. We owe a special debt of gratitude to Tim Flach for giving of his time, expertise, and remarkable talents to photograph and produce the objects for the volume.

Finally, we would like to express our sincere thanks to Tim Mennel, our editor at the University of Chicago Press, for his enthusiasm, insights, commitment, and belief in the project, and to our anonymous reviewers, who helped us to refine the conceptual framing and polish the individual fragments within this collection.

Preface

Gregg Mitman, Marco Armiero, and Robert S. Emmett

To consider humans as a geological force on Earth is to alter our very notions of time and history: geological, evolutionary, ecological, and human. As we become increasingly aware of humanity's influence upon the biophysical systems of the entire planet, we find ourselves facing an uncertain future. The idea of the Anthropocene—a term coined in 2000 by paleoecologist Eugene Stoermer and atmospheric chemist Paul Crutzen for this age of humans—has prompted scientists, artists, humanists, and social scientists to engage in new ways to understand the legacies of our species' geomorphic and biomorphic powers. Whether or not the Anthropocene becomes part of the official stratigraphic record, its advent as a scientific object has already altered how we conceptualize, imagine, and inhabit time. We have not yet specified when the new era began. The Working Group on the Anthropocene recommended 1950 as the starting date because by then radioactive elements that marked the advent of the atomic bomb were detectable across the globe. Others suggest it started thousands of years earlier, with the agricultural revolution of the Neolithic period, when the cultivation of crops, domestication of animals, and large-scale human settlements began. The orientation of history is up for grabs—as are the objects that make up history's archive, that foreshadow the future, and that will bear witness to a future past. By our objects will we know us.

In the fall of 2014 the Nelson Institute's Center for Culture, History, and Environment at the University of Wisconsin-Madison, in collaboration with the

Rachel Carson Center for Environment and Society in Munich and the Environmental Humanities Laboratory at KTH Royal Institute of Technology in Stockholm, brought together artists and anthropologists, historians and geographers, literary scholars and biologists in the playful, performative space of an "Anthropocene Slam" to shape a Cabinet of Curiosities for this new age of humans. The responsive, creative spirit of the slam invited freestyle conversation, debate, and reflection on what such a cabinet should be. What objects should it house? Which issues should it speak to? What emotions might it evoke? And what range of meanings and moral tales might it contain?

Above all, in this era of extreme hydrocarbon extraction, extreme weather, and extreme economic disparity, how might certain objects make visible the uneven interplay of economic, material, and social forces that shape the relationships among human and nonhuman beings? The Anthropocene is a narrative about space, as well as time. Its sheer scope—for example, the global scale of warming temperatures, species extinction, ocean acidification—risks obliterating the differences through which its impacts are felt by different beings, occupying different ways of life, in locales across the planet.

Slam performers dramatized, versed, and otherwise made visible the ways that planetary-scale changes become apparent and leave traces in both space and time. One participant taught the audience to fold origami passenger pigeons, a species hunted to extinction within the span of 100 years. Dozens of paper birds took flight in symbolic de-extinction. Another group poured a test slab of concrete on stage and intoned an imagined chorus for this most widely used material in our increasingly built environment. Thus the objects, images, and echoes on the slam stage evoked the sedimentary remains of humanity's impact on Earth.

In contemplating and interrogating a preemptive history of the Anthropocene and its meanings, why bring attention to *objects* when the concept invites planetary-scale thinking across eons? Neil MacGregor, in his best-selling *A History of the World in 100 Objects*, suggests that a history told through objects is a history that speaks "to whole societies and complex processes rather than individual events" (2010, xv). Just as paleontologists look to fossil remains to infer past conditions of life on earth, so might past and present-day objects offer clues to intertwined human and natural histories. The objects gathered in this book resemble more the tarots of a fortuneteller than the archeological finds of an expedition: they speak of the future. A jar of sand from a North Carolina beach, for example, opens our eyes to a multitude of natural and human forces shaping the ephemerality of barrier islands, changing property regimes, beach nourishment, and heightened storm surges. How vast the time scales, how illuminating the stories contained in just a few layered inches of sand.

Objects have the power to bridge spaces and join times. They can summon all at once the past, present, and future, blending the global and local—and thus they can disrupt linear narratives, including those about the Anthropocene. A mid-twentieth-century audio recording of a now-deceased Māori man mimicking songs of the huia, an extinct bird once endemic to New Zealand, connects disparate places and temporalities. The contemporary listener is tangled up in echoes of mimicry, memory, and extinction. Yet another object in this collection, the painting *Davies Creek Road*, simultaneously transports the viewer across the dream time of the Wiradjuri people, the deep time of anthropogenic extinction, and the imagined futures of a valley being transformed by rapacious demands for water in a warming world. All objects have the potential to contain a multitude of stories, if we use them as a way to consider multiple scales of space and time. The fuller dimensions and rhetorical weight of such object-stories can generate resistance to a narrowing of our collective possibilities. There need not be only one future, determined entirely by global climate change (Hulme 2011, 245).

Objects, too, can disrupt a sense of human exceptionalism. Such exceptionalism has sometimes been used to separate the human from nature to subdue it and foretell a future where geo-engineering might solve the planetary mess we are in. But such exceptionalist positions are not sustainable. The human species is becoming ever more implicated and entangled in the life worlds of other beings on this planet; we depend on each other for our mutual survival. Consider the feathered remains of a Canada goose, scraped off the fuselage of an Airbus 320-214 bound from New York City's LaGuardia Airport to Charlotte, North Carolina. The lives of the 155 human passengers on board hung in the balance after the bird collided with the plane. While the passengers survived, the bird remains—"snarge" is its technical term—invite us to reflect on the casualties caused directly and indirectly by accelerated lifestyles. Such bird remains ask us to contemplate the possibility of transportation infrastructures that acknowledge we share this world with diverse forms of life, all moving at different speeds and through different kinds of spaces. To see objects not just through the lens of human agency but through the lives of nonhuman beings that both shape and are shaped by relationships and processes embodied in material forms is to invite stories—in fossilized bones, decaying tissues, and living flesh. Such stories in turn bear witness to planetary-scale changes in which all species have been active participants.

Objects can also engage many publics. They can evoke inquiry, spark curiosity, and elicit tales not bound by any one discipline, language, or culture—and in so doing they can give voice to the human and even the nonhuman. Such is the case with the objects here. The voices of anthropologists and biologists, literary critics and geographers, historians and sociologists, and artists and writers are all gathered in these pages. But the voices and objects here do largely reflect perspectives from the global North. As this collection circulates the globe, what additional objects, what other tales might it stimulate?

Collectively, the objects in this book constitute a kind of Cabinet of Curiosities for the Anthropocene. Popular in the late sixteenth and seventeenth centuries, *Wunderkammern* blurred boundaries, displaying the artificial and the natural side by side (Daston and Park 1998). The marvels in them were meant to inspire a range of emotions: wonder, envy, pleasure, and fear. The Anthropocene, by also troubling boundaries between artifice and nature, can provoke similar feelings and a wide range of expressions. It has provoked utmost hubris, as in Stewart Brand's widely circulated remark, "we are as gods and have to get good at it" (2010, 20). And it has inspired more meditative, humble reflections in the face of widespread accelerated extinctions, reflected in Thom van Dooren's question: "What obligations do we have to hold open space in the world for other living beings?" (2014, 5). Technocratic optimism, ecological declension, and ethical apprehension exist side by side in future imaginaries. To collect objects *of* the Anthropocene is to register the diverse emotional responses—loss, grief, hubris, humility, anger, and pain, among others—evoked in a climate of change and uncertainty.

If there is one emotional register that unites these essays, it is curiosity—a curiosity bound intimately to care. Indeed, "caring," Donna Haraway suggests, "means becoming subject to the unsettling obligation of curiosity, which requires knowing more at the end of the day than at the beginning" (2007, 36). By drawing us outside ourselves, curiosity can shake up our place in the world. We would argue, like Vladimir Nabokov, that curiosity is insubordination in its purest form (Gade 2011, 14). Hence, reader beware: curiosity matters more than the cabinet.

Such insubordination is necessary to temper the allure of things that have the potential to reify a familiar world. Objects, as Pier Paolo Pasolini wrote, are "containers in which is stored a universe I can extract and look at." They can teach us about our place in the world. But we need also to be cautious, Pasolini warns, of the "authoritarian and repressive" character of things that can transform a limited world into a "cosmically absolute" universe. Familiar things have the potential to make other objects "extraneous, anomalous, disquieting and devoid of truth" (1987, 29–30). Objects, then, can just as easily outshine as open up other worlds. The challenge is to ask not only what objects reveal but also what they hide. We need to take notice of less familiar things, such as the goanna in *Davies Creek Road*. These entertain the possibility of other beings, other relations in the world, and other cosmologies not easily subsumed within the dominant tropes of Western science animated by one version of the Anthropocene—as a fable of civilizational progress.

Tim Flach's photographs of the objects found in our Cabinet of Curiosities for the Anthropocene are also driven by a sense of intrigue and curiosity, inviting the viewer to imagine and explore the past, present, and potentially future meanings of these fossils. Flach's images suggest each thing's characteristic trait: the brutality of concrete, the forensic nature of a feather, the extinct form of a Blackberry. A London-based photographer whose animal images circulate around the world and provoke questions of what it means to be human, Flach brings to this project an aesthetic sensibility, keen understanding, and technical brilliance in creating wondrous images in the spirit of a Cabinet of Curiosities.

In these strange and uncertain times, the curious juxtapositions of *Wunderkammern*, as Libby Robin argues here, invite a salutary reconsideration of the Enlightenment notion of a humanity set apart from Nature that has held sway even as it has become apparent that we live in a post-natural world. The objects in this cabinet join that long-term work of uniting art and science, natural and unnatural histories, and enlivening new makers and publics to respond to the planetary impact of human activities. This volume is less a catalog than a series of reflective essays organized around fifteen exemplary objects that offer a fragmentary history of the Anthropocene. Its curated selection of "remains" calls for readers to browse, dip in, and explore. Instead of providing a single overarching narrative—whether of a negative universal history of humanity's ecological destruction or a triumphal prediction of a bright and perfectly engineered future—these remains interrogate the limits of the idea of the Anthropocene and make us wonder anew about what human history is made of.

BIBLIOGRAPHY

Brand, S. 2010. *Whole Earth Discipline*. New York: Penguin Books.
Daston, L., and K. Park. 1998. *Wonders and the Order of Nature, 1150–1750*. New York: Zone Books / MIT Press.
Gade, D. 2011. *Curiosity, Inquiry, and the Geographical Imagination*. Bern: Peter Lang.
Haraway, D. 2007. *When Species Meet*. Minneapolis: University of Minnesota Press.

Hulme, M. 2011. "Reducing the Future to Climate: A Story of Climate Determinism and Reductionism." *Osiris* 26 (1): 245.

MacGregor, N. 2010. *A History of the World in 100 Objects*. New York: Viking.

Pasolini, P. P. 1987. *Lutheran Letters*. Translated by Stuart Hood. New York: Carcanet.

Van Dooren, T. 2014. *Flight Ways: Life and Loss at the Edge of Extinction*. New York, Columbia University Press.

The Anthropocene

The Promise and Pitfalls of an Epochal Idea

Rob Nixon

Time gets thicker, light gets dim
    Allen Ginsberg, "The Gates of Wrath"

What does it mean to imagine *Homo sapiens* as not merely a historical but a geological actor, a force of such magnitude that our impacts are being written into the fossil record? What does it mean to acknowledge that, for the first time in Earth's history, a sentient species, our own, has shaken Earth's life systems with a profundity that paleontologist Anthony Barnosky has likened to an asteroid strike? How does that perceptual shift disturb widespread assumptions about the deep past and the far future, about planetary history, human power relations, and the dynamics between humans and nonhuman agents of Earth's metamorphosis? If our actions have become geologically consequential, leaving an enduring archive that will be legible for tens or even hundreds of millennia to come, what will that archive disclose about social relations, above all, about the unequal weight of human communities possessing disparate earth-changing powers? And, in terms of the history of ideas, why now? Why has the idea of *Homo sapiens* as a fused biological-geological force gained traction in the second decade of the twenty-first century, when in the twentieth century geologists typically dismissed our species' occupancy of this planet as not just ephemeral but as geologically trivial?

Such consequential questions follow from the turn to the Anthropocene, a hypothesis advanced by Nobel Prize–winning atmospheric chemist Paul Crutzen and paleoecologist Eugene Stoermer in 2000. Stoermer had been using the term "Anthropocene" informally since the 1980s, but it only achieved academic prominence when the Nobel Prize–winning Crutzen threw his weight behind it and, together with Stoermer, gave the term an interdisciplinary reach and urgency. Crutzen and Stoermer argued that the Holocene was history: the earth had entered a new, unprecedented geological epoch, triggered by human actions. The Anthropocene has many disputed beginnings: some date its emergence to the rise of sedentary agricultural communities roughly 12,000 years ago, others to 1610 and the colonization of the Americas, others still to the onset of Europe's industrial revolution circa 1800 or to the Trinity nuclear test of 1945.

Crutzen and Stoermer favored placing the golden spike—locating the Anthropocene break—in the late eighteenth-century beginnings of the Industrial Revolution, and this remains the most broadly cited position. According to their dominant Anthropocene script, over the past two and a quarter centuries we have been laying down in stone a durable archive of human impacts to Earth's geophysical and biophysical systems. Those long-term impacts have become particularly acute since 1945 during the so-called Great Acceleration. We have decisively altered the carbon cycle, the nitrogen cycle, and the rate of extinction. We have created unprecedented radionuclides and fossilized plastics. We have erected megacities that will leave an enduring footprint long after they have ceased to function as cities. We have changed the pH of the oceans and have shunted so many life forms around the globe—inadvertently and intentionally—that we are creating novel ecosystems everywhere. Of vertebrate terrestrial life, humans and our domesticated animals now constitute over 90 percent by weight, with less than 10 percent comprised by wild creatures. Over the past century the global temperature has risen ten times faster than the average rate of Ice Age–recovery warming. Over the next century that rate is predicted to accelerate at twenty times the average. What kinds of signals will all these momentous changes leave in the fossil record?

The Anthropocene's Interdisciplinary Energy

When Crutzen and Stoermer (2000) advanced their hypothesis, they couldn't possibly have imagined what an immense, omnivorous idea it would become. It took a while, but by the millennium's second decade those enthralled and appalled by the Anthropocene were being sucked, in their interdisciplinary masses, into its cavernous maw. Enthusiasts and skeptics poured in from paleobotany and postcolonial studies, from nanotechnology and bioethics, from Egyptology, evolutionary robotics, feminist psychology, geophysics, agronomy, posthumanism, and druidic studies. The classicists arrived alongside the futurists, where they mingled with students of everything from plastiglomerates to romantic prosody, from ruins to rewilding.

This has arguably been the most generative feature of the Anthropocene turn: the myriad exchanges it has stimulated across the earth and life sciences, the social sciences, the humanities and the arts, bringing into conversation scholars who have been lured out of their specialist bubbles to engage energetically with unfamiliar interlocutors. The Anthropocene, at its best, has prompted forms of interdisciplinary exchange that didn't exist before, giving impetus to creative collaborations across intellectually debilitating—dare one say fossilized—divides. Despite some of the nefarious uses to which it has been put, the Anthropocene paradigm can be used productively to pose large questions about the ways we partition knowledge and delimit being.

The humanities and arts have become vital to the conversational mix over what the Anthropocene can and should convey, which is as it should be. For the Anthropocene—or at least the iconoclastic part of it—began as a provocation, an exhortation, a shock strategy of a kind that we are attuned to in the arts and the humanities. What will the world look like if you change the frame, scramble the view, upend the perspective, in pursuit of some startled state of sensory and imaginative vitality? A quest for creative disturbance is one impulse behind the Cabinet of Curiosities, which gives body to a conviction that rarefied theorizing needs to be grounded in intimate encounters. For there is a real risk that the Anthropocene at its most compendious can be diminishing, promulgating—ironically, for a theory of expanded human agency—a mood of inaction, quietism, nihilism, inertia.

To give any version of the Anthropocene a public resonance involves choosing objects, images, and stories that will make visceral those tumultuous geologic processes that now happen on human time scales. The lively array of object-driven stories assembled for the Cabinet of Curiosities affords immense biomorphic and geomorphic changes a granular intimacy. Encounters with the granular—as opposed to the grandiose—world, can, depending on one's perspective, conceal or reveal. Imaginative revelations may prompt modest moments of self-transformation, but they need not be limited to that, as we have seen in the ever more dynamic relations emerging between the visual arts, the performing arts, and the climate justice movement, a dynamic that has helped shift political and ethical sightlines. Above all, to insist on the value of imaginative encounter—be it with a fossilized Blackberry, a cryogenic zoo, a jar of sand, a cement mixer, or the lonely mating call of an extinct bird—is to refuse the quantifiers ownership of the Anthropocene, to insist that the immeasurable power of storytelling and image making is irreducible to the metrics of human impacts. Indeed, the arts and humanities can serve a restraining order on the runaway hubris of technocratic Anthropocene expertise by resisting the political logic of Team Future, whereby those who crunch the numbers are first in line to engineer the new worlds.

If the Anthropocene is reverberating across the humanities, this makes another kind of sense, for it shakes the very idea of what it means to be human. To invest a young species like Homo sapiens with geologic powers—to open up the human to what in the postenlightenment would be considered inhuman time scales—is a tectonic act. We're simply not accustomed, maybe even equipped, to conceive of human consequences across such a vastly expanded temporal stage, across which we stride as (more or less) ambulatory rocks. To revisit Barnosky's asteroid trope, what does it mean for the "being" in "human being" to depict us as a hurtling hunk of rock that feels?

The novelist Amitav Ghosh, in a series of perceptive lectures, has suggested how the Anthropocene turn can help us recognize the imaginative limits of the forms—from the arts to urban planning—favored by enlightenment modernity. Ghosh (2016) observes how the legacy of enlightenment modernity's attachment to linear progress has suppressed modernization's contradictions, hindering the imaginative and strategic responses to the Anthropocene and the global climate crash. The realist novel that fed off and advanced an idea of linear progress typically centered on a small cast of characters and a delimited landscape that became background to the unfolding action. But the Anthropocene has made the environment as background to the growth of character untenable, as it becomes increasingly difficult to ignore the inconceivably vast forces emanating from the environment, forces entangled with human actions but scarcely subordinate to them. The realist novel, in contrast to a form like the epic, has proven ill-equipped to make the vast scalar leaps across space and time that the Anthropocene demands, leaps from the cosmological to the microbial, from the deep past to the remote future. Moreover, the design of enlightenment forms like the realist novel and the colonial city downplayed the irruptive powers of nonhuman actors: the unruliness of volcanoes, rivers, locusts, rats, shape-shifting leopards, and moody mountains, all of which in the epic speak to the arrogant limits of an isolationist view of human development.

Rational enlightenment forms like the realist novel and the colonial city, Ghosh suggests, have suppressed vital intuitions about the vulnerability of human designs to forces that other art forms and other cosmologies have kept alive through an awareness of human precariousness before the powers exercised (for good and ill) by nonhuman actors. Indeed, the refusal of the human-nonhuman distinction—by now such a central theme of Anthropocene thought—has persisted in many cultures in a state of contradictory entanglement with developmental modernity. Could the rise of animal studies be linked in this way to climate chaos, to a disillusionment with a separationist, hubristic ideology of hyperationality, and to a renewed fascination with the instinctual, the bodily, the ineluctable connectedness between us and the biota that permeate our lives? And could it be, as Ghosh argues in a suggestion of direct pertinence to the Cabinet of Curiosities, that digital culture's reassertion of imagistic power over the enlightenment's elevation of the word has created a hybridized image-word milieu that is more responsive to the challenges of Anthropocene representation than the word-besotted, linear forms that the enlightenment extolled?

The imaginative questions that the Anthropocene provokes are accompanied by historical ones. The Anthropocene has profound implications for the meaning and object of history, reframing the future by rethinking the past as shaped by a fused biological-geological actor. Crutzen and Stoermer's neologism is both historically belated—suggesting that people possessed planetary geomorphic powers long before they realized it—and anticipatory. For if our actions have indeed propelled us beyond the Holocene, the new epoch we have set in motion is in its infancy. The Anthropocene thus pulls us simultaneously into deep pasts and deep futures that are unfamiliar, uncomfortable terrain for historiography.

The implications of the Anthropocene for history making are inseparable from the history of technology. New technologies of detection have generated new geophysical archives of inquiry that are reshaping—across the sciences, the social sciences, the humanities, and the arts—assumptions about what stored knowledge looks like, about archival reading practices, and about the interdisciplinary literacy such readings may require. The advent of paleoclimatology and dendroclimatology, our ability to posit tree growth rings, ice cores, deep sea cores, and fossil soils as proxies for past climates, the rise and spread of drones, and ever more elaborate satellite imaging all allow us to generate more varied perspectives, newly minute and newly vast, on planetary life and time.

But if new technologies of detection have proven crucial to the Anthropocene's burgeoning authority, the technological dimension can mask relations of power. Who gets to don the white coat of expertise? Who becomes central, and who marginal, in the contest over narrative authority? As Susan Schuppli (2014) observes in her work on material witnessing, traces of the apparently inanimate world can be given voice by increasingly sophisticated technologies. But there is inevitably conflict over what stories those material traces release in, for example, a war tribunal or a truth commission. Who gets to dragoon those traces into delivering certain kinds of stories as opposed to others? Such questions pertain with equal force to the contouring of the Anthropocene grand narrative. From the perspectives of anticolonialism, feminism, multispecies ethnography, queer ecologies, and environmental justice, among others, we are seeing the emergence of a kind of strategic witnessing, a pushback against the risk that the Anthropocene may become a resurrected selective enlightenment in disguise, an apparently novel but potentially regressive Age of Man.

Anthropocene Pitfalls

To gauge the promise and pitfalls of the Anthropocene we need to position the proposed epoch in the history of ideas. As has been noted, Crutzen and Stoermer's theory had several partial precursors. But there is a more recent history that has been overlooked. Crutzen and Stoermer began promoting the Anthropocene together in 2000, but for almost ten years it achieved very little public resonance. The debates over the merits of the term were rarely heard outside narrow intellectual corridors, dominated by a handful of earth scientists, life scientists, and archaeologists. How do we explain the belated emergence of a more public Anthropocene? How do we explain the paradigm's lost decade?

Less than a year after Crutzen and Stoermer launched their explosive vision of humanity as geological actor, 9/11 happened. Then in 2002 the Bali bombings killed 202 people (Australian tourists comprising the largest number), followed by the Bush-Blair invasion of Iraq in 2003, the 2004 Madrid train bombings, and the 2005 bombings in London. Of course, greater numbers of people were killed elsewhere—by state and nonstate actors—but those bombings were the ones that most viscerally shook Westerners' faith in history's continuity, catapulting them into a feeling that "people like us" had entered a new age of violent vulnerability. Against this backdrop, time shrunk. And the efforts of an atmospheric chemist and a paleoecologist to expand time—or, metaphorically, to explode our temporal norms—was no match for the bomb-dominated temporal frameworks of the day. The vast scales of geologic time, even the more modest intergenerational times scales of accelerated climate change, were inimical to the dominant perceptions of catastrophe. In a news cycle fixated even more than usual on instantaneous violence, a preoccupation with Islamic extremism marginalized efforts to dramatize how extreme climate change and extreme extractive practices (tar sands, cold-water deep-sea drilling) would incrementally inflict untold human and ecological casualties. The 2008 Great Recession reinforced this bias toward instant crisis, especially in the United States, where Big Carbon bankrolled the zero-sum ideology of jobs versus the environment as part of its perpetual war on climate science. In short, during the millennium's first decade, both the long emergency of the climate crisis and the even longer emergency of the Anthropocene struggled to gain urgency in an inhospitable political and temporal frame.

In the millennium's second decade the Anthropocene has begun to spread beyond the university and permeate the public sphere. Bloggers, filmmakers, public intellectuals, and curators are now trying to reimagine, through the prism of the Anthropocene, what geographer Doreen Massey calls "the ancient manoeverings of life and rock" (2005). Debates over the paradigm's merits and implications have attracted an ever-wider cast of disciplines and arts. We have seen special Anthropocene issues or cover stories in the *Economist*, *Nature*, *National Geographic*, and *Smithsonian* and lively debates hosted by the *New York Times* and the BBC. From Germany to Australia, Switzerland to the United States, curators are staging ambitious Anthropocene shows that range, in mood, from the celebratory to the despairing, from the earnest to the antic. The Age of the Human is making itself felt in modest galleries and mega art shows, from the Venice Biennale to Art Basel Miami. The Anthropocene's digital presence has also skyrocketed on Flickr, YouTube and in Ted Talks. A Google Alert that yielded five results a week in 2011 yielded seventy a week four years later.

Yet the timing of the Anthropocene's breakthrough into the public realm coincided with another public turn in the history of ideas. The millennium's second decade also saw an even more decisive rise in public attention to an apparently unrelated issue: deepening economic inequality, in society after society—in countries as varied as China, Sweden, South Africa, Argentina, Italy, Jamaica, the United States, India, Nigeria, Indonesia, and the United Kingdom.

The disparities are alarming. In 1980 the average American worker-to-CEO pay ratio was 1:40. By 2014 that ratio had soared to 1:296. New York City boasts seventy billionaires, yet 30 percent of the city's children languish in poverty. In a single year, 2013, the average price of a London home soared by $120,000. In South Africa the two wealthiest businessmen (both white) have amassed a net worth that surpasses the net worth of the nation's poorest 50 percent. Californians burn more gasoline than the 900 million inhabitants of Africa's fifty-four nations combined. A one-way flight from Los Angeles to New York produces more carbon emissions than the average Nigerian does annually. Oxfam reports that in 2013 the combined wealth of the world's richest eighty-five individuals equaled that of the 3.5 billion people who constitute the poorest half of the planet. And a 2013 study concluded that since 1751—a period that encompasses the entire Anthropocene to date—a mere ninety corporations have been responsible for two-thirds of humanity's greenhouse gas emissions. That's an extraordinary concentration of earth-altering power.

With few exceptions, discussions of the Anthropocene and inequality have tended to travel along parallel paths. Yet what does it mean, in terms of the history of ideas, that the Anthropocene as a grand explanatory species story has taken hold in plutocratic times, when economic, social, and environmental injustice is marked by a deepening schism between the uber-rich and the ultra-poor, between gated resource-hogs and the abandoned destitute? Doesn't lumping together under the sign of the human the average twenty-first-century Liberian and the average twenty-first-century American as agents of planetary change risk concealing more than it reveals? Is it not the case, as Jennifer Jacquet (2013) has suggested, that while some humans are leaving Anthropocene footprints that are indubitably geological, other humans are not geological actors at all? There is of course a profound need for concerted action to slow the most deleterious, life-threatening processes of anthropogenic planetary change in order to secure viable futures. But the call for coordinated transnational strategies should not become the kind of totalizing gesture that suppresses the radically unequal history of human impacts and hence of human responsibilities.

Imaginative perspectives have political implications. An epic Anthropocene vantage point risks concealing—historically and in the present—unequal human impacts, unequal human agency, and unequal human vulnerabilities. So a crucial challenge facing us is this: how do we tell two large stories that can often seem in tension with each other, a convergent story and a divergent one? First, a collective story about humanity's impacts that will be legible in the earth's geophysical systems for millennia to come. Second, a much more fractured story that acknowledges dramatic disparities in planet-altering powers. For Anthropocene thinking to retain any credibility, it needs to negotiate the complex dynamic between a shared geomorphic narrative and increasingly unshared resources. We may all be in the Anthropocene but we're not all in it in the same way.

"We" is a tricky word at the best of times, doubly so in the context of Anthropocene-species speak where "we" serves as an assumed point of departure, not the product of historical contingencies. Stylistically, "we" is difficult to avoid unless the writer ducks behind the passive voice, that hiding place of preference for academics determined to avoid confronting the subject of agency head on. In Anthropocene thought, agency becomes a particularly high-stakes game, as evolutionary psychologists, neurobiologists, new geologists, philosophers, and liberal humanists emit mutually reinforcing "we's" that are too often deficient in any textured acknowledgment that "we" is a historically and culturally shape-shifting formation. There is no transcendent "we," Anthropocene or otherwise: the appearance and disappearance of collective identities is inseparable from the vexed institutional histories that contour struggles over power.

If public attention to inequality has risen during the millennium's second decade, it has been assisted by the spread of the 1 percent meme that Occupy coined and—alongside the so-called international square movements—helped disseminate. However, the narrative tension between a unitary species narrative and socioeconomic fracture does not exist merely in relation to current practices, but reaches back into industrial, colonial, and neoliberal history. Indeed, if the mid-twentieth century marks, in most Anthropocene accounts, the advent of the Great Acceleration in human impacts, the bulk of the period since has been distinguished by the spread of neoliberal practices through the Washington Consensus, the World Bank, the IMF, the rise of the antiregulatory World Trade Organization and the Reagan-Thatcher counterrevolution, practices that have accelerated the globalization of elite resource capture. Despite determined resistance to neoliberalism, we have witnessed increasing attacks on the public sphere and a retreat, across many societies, from governmental responsibility for citizens' basic welfare amid what George Monbiot calls "a bonfire of regulation" (2010).

In 1987 Margaret Thatcher notoriously declared, "There is no such thing as society. There are individual men and women and there are families." Less than a year later, James Hansen (then director of NASA's Goddard Institute) delivered a historic address calling for collective action to avert climate change. Hansen testified before US congressional hearings that climate science was 99 percent unequivocal: the world was warming and we needed to act collaboratively to reduce emissions. So just as Hansen was summoning humanity to tackle collectively a problem too vast to be fixed by individual lightbulb-changing efforts, the very idea of collective identity and collective values, indeed, the very idea of the public, was being ridiculed and assailed. Such assaults helped promulgate an ideology of hyperindividualism and hyperconsumption that twinned freedom to atomized consumer choice and vilified government as freedom's adversary. Thus the idea of the public good atrophied in favor of individual consumer goods, resulting in a scaled-down civic sphere mismatched to a scaled-up climate crisis. A similar mismatch emerged in the domain of environmental time: a crisis that demanded collaboration for long-term collective survival was ill-fitted to the accelerated, hypercarbonized pursuit of immediate wealth at any cost by megacorporations unanswerable to the longue durée, corporations that became more mobile, wealthier, and more powerful than most of the societies they operated in. All this has had profound implications for environmental justice during the Great Acceleration, as the crisis of futurity has become inextricable from the neoliberal crisis of disparity.

So any account of the political, ethical, and narrative challenges inherent in the Anthropocene needs to address the relationship between the Great Acceleration and the Great Divide, the economic splintering under neoliberalism that Timothy Noah (2013) has also called the Great Divergence. Diane Ackerman's best-selling *The Human Age* (2015), one of the most influential public explorations of the Anthropocene, demonstrates the costs of ignoring the connections between the Great Acceleration and the Great Divide. During her sunny-side up tour of Anthropocene effects, she encounters the futurist Ray Kurzweil, whom she quotes uncritically as predicting that "by the 2030s we'll be putting millions of nanobots inside our bodies to augment our immune system, to basically wipe out disease." Pray tell, which "we" would that be?

Technological innovation will clearly play a critical part in the battle to adapt to the breakneck pace of anthropogenic planetary change, but let's acknowledge that we're doing a far better job of encouraging innovation than distributing possibility. One billion people remain chronically hungry, while 2.5 billion survive on less than two dollars a day. According to the United Nations Refugee Agency, in 2014 the number of displaced people reached 59.5 million, the highest figure ever recorded. One out of every 122 humans is now either a refugee, an asylum seeker, or internally displaced. As the report observes: "If this were the population of a country, it would be the world's 24th biggest."

In celebrating the culture of innovation, Ackerman focuses on the interplay between technology, design, and rapid evolution. But what of the decisive role played by forms of governance? In the plutocratic milieu of the twenty-first century, how do we ensure that innovations aren't by the few for the few, that they don't compound the trend toward islands of extreme affluence barricaded against vulnerable multitudes?

A technology's emergence is no guarantee that its benefits will trickle down to humanity at large. When men gang-raped two teenage girls and hanged them from mango trees in India in 2014, the atrocity drew attention to the fact that the women had to risk entering the forest at night in order to defecate. Two and a half billion humans still lack access to a rudimentary latrine, a venerable technology developed 5,000 years ago. Deprived of any formal sanitation, residents of Kibera, the sprawling Nairobi slum, routinely resort to "flying toilets," defecating into plastic bags that they hurl onto the streets. Indeed, the absence of sanitation—and consequent water pollution—causes, by some calculations, 70 percent of diseases globally.

Can Anthropocene-inspired thought help generate more equitable policies in a spirit of transformative justice? Can the Anthropocene help rouse citizens and governments to act for long-term, concerted change? Those are vast questions, but they remain essential ones. Too often the Anthropocene assumes a hasty universalism that masks the connection between our conjoined crises—between accelerating environmental devastation and rising inequality. As Andreas Malm notes, "Dehistoricizing, universalizing, eternalizing, and naturalizing a mode of production specific to a certain time and place—these are the classic strategies of ideological legitimation" (Malm and Hornborg 2014). To these strategies we can now add "geologizing" as a way of legitimating processes that could proceed very differently under more progressive, more equitable economic systems and forms of governance.

Despite efforts to communicate Anthropocene thought to nonspecialist audiences, concerns linger over the limited demographic character of the paradigm's public appeal. The Anthropocene remains a heavily top-down model: the rethinking from above is unmet by any comparable energies rising from below. By contrast, we have witnessed a truly diverse, international cast of communities mobilize behind calls for environmental justice, climate justice, indigenous rights, and the environmentalism of the poor. We have also witnessed communities mobilize internationally against neoliberalism and austerity. In such contexts, memes like the 1 percent and "We Can't Breathe" have rippled well beyond their origins. But we have yet to witness any similar spread of the Anthropocene —and I suspect are unlikely to—as an international mobilizing device. The term has several disadvantages stacked against it. For one thing, it sounds academic: there's an arcane, egghead quality to the word. Unlike, say, indigenous rights or the environmentalism of the poor, "the Anthropocene" is not self-explanatory, but requires elaborate intellectual mediation.

The intellectual mediators have been mostly white and male, in a process that often has an unselfconsciously regressive dimension. Two European men first advanced the Anthropocene thesis and the trajectory of its spread has failed to shake that association. Anthropocene theorists, mediators, and popularizers remain overwhelmingly white Europeans, North Americans, and Australians and skew heavily male. In Africa, Asia, and Latin America the Anthropocene has failed to achieve any comparable purchase among thought leaders, let alone any resonance on the streets. The situation is familiar: the few talking among themselves—as if their confined demographic were universal—on behalf of the many.

Such a dynamic is perturbing, doubly so because "on behalf of " are loaded words in the context of the Anthropocene's fraught politics of agency. As the only inhabitable planet is losing that vital quality in irreversible increments, who will get to be the deciders? Who will rearrange the conditions of life for all humanity's social strata and for other organisms as well? Will the architects of those rearrangements naively and dangerously assume that humans are the planet's only consequential actors? If humanity—or rather, a selective set of humanity—has narrowed the possible trajectories of Earth's future, who will determine the necessary, always imperfect, countercorrections? Will the drivers of change be swayed by lobbyists, enjoy corporate sponsorship, and stack their think-tanks with billionaire philanthropists who speak in visions of innovation, sustainability, resilience, and adaptation, those increasingly green-washed, greed-tainted words, retrofitted to neoliberal policies?

We face the dismaying prospect that the Anthropocene will be mobilized in increasingly autocratic ways that flow from—and potentially exacerbate—the authoritarian streak already evident in neoliberal practices. To allow plutocrats to deputize for the species would represent a new twist in the sorry history of government for the people without the people.

Species thinking, particularly when partnered with Silicon Valley–style technoexuberance, tends to sidestep thorny questions of representative governance. That tendency is evident in those we might call command-and-control Anthropocene optimists, like ecologist Erle Ellis, who believes "we must not see the Anthropocene as a crisis, but as the beginning of a new geological epoch ripe with human-directed opportunity" (2011). Ranked alongside him are science journalists Mark Lynas (author of *The God Species*) and Ronald Bailey, who insists that "over time, we will only get better at being the guardian gods of the earth" (Bailey 2011). As their mantra, these Anthropocene optimists cite Stewart Brand's exhortation: "we are as gods and must get good at it" (Brockman 2009).

But for others, talk of *Homo sapiens* as god species, as Earth's surrogate divinity, is positively chilling. Hasn't a hubristic mindset of earth mastery, of dominion over nature, gotten us into this mess as out-of-control geological actors? Earth mastery, moreover, conjures up disturbing associations with the race, gender, and class hierarchies of the selective enlightenment. More than twenty years ago, feminist scholars like Anne McClintock (1995) and Val Plumwood (1994) were laying bare the implications of the standpoint of mastery. The climatologist Mike

Hulme (2014) sees a direct line between the new wave of magisterial thinking and the reckless adventurism of a small, powerful set of geo-engineers and their billionaire backers who harbor ambitions to "reset the global thermostat." Philanthrocapitalists with designs on "fixing" the global climate now sponsor research with that end in mind at a variety of think-tanks and universities.

We should not equate human planetary impact with human planetary control, as either a possibility or an ideal. Although Crutzen initially floated the possibility of climate engineering, he later backed away from that intimation. Giddy fantasies of omnipotence are a far cry from the stronger, cautionary impulse that animates his work: "what I hope is that the term Anthropocene will be a warning to the world" (Kolbert 2011). In heeding that warning, we need to face the incalculable complexities of a rapidly changing Earth by shedding illusions of mastery and adopting instead an engaged humility that is not synonymous with quietism.

The Breakthrough Institute has become the primary think-tank for Anthropocene brightsiders, the self-declared ecomodernists. The Ecomodernist Manifesto sets out to "use humanity's extraordinary powers in service of creating a good Anthropocene." The institute's Ted Nordhaus and Michael Shellenberger (2009) berate environmentalists who "warn that degrading nonhuman natures will undermine the basis for human civilization. But history has shown the opposite: the degradation of nonhuman environments has made us rich." But as Chris Smaje (2015) notes: "There is no sense [in the Ecomodernist Manifesto] that processes of modernisation cause any poverty. . . . There's nothing on uneven development, historical cores and peripheries, proletarianisation, colonial land appropriation and the implications of all this for social equality. The ecomodernist solution to poverty is simply more modernization." To embrace the kind of uncritical techno-idealism that the Breakthrough Institute promulgates is to gloss over the violent, socially divisive history of environmentally unsustainable practices that designate certain communities and ecosystems as disposable in the name of the modernity's onward march. In keeping with the long, suspect history of modernization theory, that march becomes an innate good. It does so regardless of whether it exacerbates brutally exclusive practices of resource capture, human and nonhuman community abandonment, and the creation of uninhabitable sacrifice zones.

Science writer Elizabeth Kolbert has tweeted: "two words that probably should not be used in sequence: 'good' & 'anthropocene'" (2014). Environmental philosopher Kathleen Dean Moore goes further, suggesting that the Anthropocene would have been better named the Unforgiveable-crimescene (2013). Nonetheless, the technocratic hubris of ecomodernist thinking is powerfully inflecting the way the Anthropocene is being activated in the public sphere. Here's Ackerman's celebration of the new epoch: "We are dreamsmiths and wonder-workers. What a marvel we've become, a species with planet-wide powers and breathtaking gifts" (2015, 310). That we may be, but such awestruck Anthropocene optimism can feel eerily unearned in the absence of a measured acknowledgment of the losses, the traumas, the scars that afflict the mesh of human and nonhuman communities in this volatile new epoch. And so the ecomodernists become the grief police: no mourning permitted here, move on already, you're creating an inadmissible disturbance.

Command-and-control Anthropocene thinking evidences other limitations. Does not calling something the Age of Humans risk an isolationist mentality that encourages species narcissism? It's one thing to recognize that *Homo sapiens* has accrued massive bio- and geomorphic powers. But it's another thing altogether to fixate on human agency to a degree that downplays the imperfectly understood, infinitely elaborate matrices of nonhuman agency, from the microbiome to the movement of tectonic plates, that continue to shape Earth's life systems. To be sure, humans—especially the wealthiest among us—possess planet-altering powers, but we do not exercise those powers in a state of segregation from the actions of other forces. As Aldo Leopold noted many decades ago, dreams of environmental mastery are nothing but "biotic arrogance" (1935).

Ecomodernists tend to posit humanity in the aggregate as bossing the biosphere, as the indisputable winner in the planet-altering stakes. But that assumption exaggerates the cohesiveness of "the human" actor while simultaneously ignoring the earth-altering effects that flow from interspecies actions, be they collaborative or competitive. To treat *Homo sapiens* as a transcendent super species is to head down the slippery slope of exceptionalism. For the ecomodernists, it is a short, untroubled step from human exceptionalism to technocratic exceptionalism, as a small, unelected clique of visionary "innovators" gets tasked with leading the species to higher ground. But what are the geopolitics of this synecdoche? By what right do the technocrats elevate themselves as humanity's self-annointed "mini-we," in a time of heightened economic fracturing, of wildly disparate levels of vulnerability, when the burden of resilience and the possibilities of survival are marked by brutal disparities?

At least pause to ponder this: is it ethical that as the super-rich capture ever more resources, the outcast poor, who have contributed least to our planet's headlong transformation, are abandoned to the climate frontlines where they must weather the brunt of the chaotic effects? Ackerman may be right that "a warmer world won't be terrible for everyone, and is bound to inspire new technologies

and good surprises, not just tragedy" (2015). But her assertion deserves a follow
up question: who is in line for the good surprises and who is queuing up for tragedy? Hurricane Sandy brought precisely that question to the fore. Manhattan?
Too valuable to lose. Bangladesh, even Far Rockaway, not so much.

Conclusion

In the annals of the Anthropocene, August 29, 2016, proved to be a significant—if
not yet decisive—date. On that day the Anthropocene Working Group, an interdisciplinary and international committee of geoscientists, recommended to the
International Geological Congress gathered in Cape Town that the Anthropocene
be formalized as a new epoch within the Geologic Time Scale. It took the group
seven years to sift the evidence and debate the science before they concluded that
the geological signals were sufficiently strong and incontrovertible to warrant recognizing the new epoch, which began, they argued, circa 1950. However, even
this laboriously achieved consensus far from guarantees that the Anthropocene
will gain official acceptance. The deliberations of the International Commission
of Stratigraphy and the International Union of Geological Sciences must now inch
forward as they in turn weigh the merits of the working group's recommendation.

But how many people are waiting, with bated breath, for a tiny circle of geological deciders to certify the Anthropocene's existence or otherwise? If the
deciders determine that the Anthropocene is nothing but a deep-time hallucina-
tion, will the idea vanish from museums and art galleries, from rich-nation scholarly debates and interdisciplinary conventions, from Flickr and YouTube? Not
likely. Ironically, the geologists' slow-motion—dare one say glacial—assessment
of the Anthropocene's claims has eroded their authority over the outcome, as in
a time of digital acceleration the paradigm has undergone a high-speed adventure through fields well beyond the jurisdiction of earth scientists. If the Anthropocene peters out, it will be through paradigmatic exhaustion and a failure to
strengthen its social purchase, not because the idea lacks geological certification.

Indeed, the Anthropocene has become so ductile, so infinitely malleable that
we should perhaps view it less as a paradigm than as a spectrum of paradigms that
range from the hubristic to the humble, from the reactionary to the positively iconoclastic. What began as a data-driven scientific debate over how to calibrate and
classify the human fingerprint in the fossil record has since spread to almost every
imaginable scholarly field, across the arts, and spilled out into the world beyond.
The Anthropocene—by turns enlightening, exasperating, alarming—throws up

questions that metrics in isolation cannot answer. As environmental historian Libby Robin observes, "the question is how people can take responsibility for and respond to their changed world. And the answer is not simply scientific and technological, but also social, cultural, political and ecological" (2008).

At its most suggestive, the Anthropocene can stimulate new forms of noticing that may help provoke layered thinking about responsibility. At its most suggestive, the Anthropocene can help us rediscover the vitality of mundane objects and, in a spirit of anticipatory memory, encourage us to grapple with the barely comprehensible, emergent worlds toward which we're plummeting. At its most suggestive, the Anthropocene can lead us toward consequential questions about the relationship between the imaginable and the unimaginable, between possible lives and probable ones, and stimulate debate over how we negotiate—from our diverse Anthropocene positions—the challenges that shadow the path ahead.

The Cabinet of Curiosities is expressive of this ambition. It can hopefully help, in some modest way, by offering alternative forms of thinking through time, alternatives to the catastrophic temporal parochialism that afflicts the neoliberal order. Collectively, the Anthropocene objects, performances, and stories arrayed here have the power to disturb and to surprise, goading us toward new ways of thinking and feeling about the planet we have inherited and the planet we will bequeath.

Yet in the context of the Anthropocene every "we" remains an uneasy one. Does "we" promulgate illusions of an isolationist human supremacy or of *Homo sapiens* as a collective actor? "Nature no longer runs the Earth," writes Mark Lynas. "We do. It is our choice what happens here" (2011). That's dubious on two fronts. First, humans are inseparable from other planet-shaping powers that are never fully other, not least microorganisms that vigorously impact the conditions of human and planetary being. Second, the notion that we "run the Earth" smacks of egotistical arrogance and suppresses the deep, painful divides in what it means to be human in a world littered with neoliberalism's sacrifice zones, a world where the ghosts of disposable people wander the perimeters of gated communities that deny them their humanity.

This, then, is surely the primary obligation that adheres to Anthropocene endeavors: to resist the imposition from above of a quick-and-easy "we" that becomes complicit in disenfranchising the many. Above all, the Anthropocene challenges us to devise more just institutions of governance that can better distribute finite resources and technologies, thereby enhancing the life chances of all humanity as well as enhancing ecosystem viability. We cannot risk giving a free pass to buccaneer billionaires, disaster profiteers, and venture philanthropists to deputize for the species. We cannot risk allowing them to usurp authority over the worlds to come, to determine who gets a future and who is denied one by geoengineering quality of life—however temporary—for elite zip codes only. We live in times that call for concerted action but only if such action acknowledges and seeks to heal the disconcerting fractures in the meaning, the possibility of the human.

BIBLIOGRAPHY

Ackerman, D. 2015. *The Human Age: The World Shaped by Us*. New York: Norton.

Anon. 2012. "Leaving Our Mark: What Will Be Left of Our Cities?" BBC News. http://www.bbc .com/news/science-environment-20154030 (accessed on February 2, 2015).

Bailey, R. 2011. "Better Be Potent Than Not." *New York Times*, May 19, 2011.

Brockman, J. 2009. "We Are as Gods and Have to Get Good at It: A Talk with Stewart Brand." *Edge*, August 18. https://www.edge.org/conversation/stewart_brand-we-are-as-gods-and -have-to-get-good-at-it (accessed October 14, 2016).

Carrington, D. 2016. "The Anthropocene Epoch: Scientists Declare Dawn of Human-influenced Age." *The Guardian*, August 29.

Crutzen, P., and E. Stoermer. 2000. "The Anthropocene." *IGBP Newsletter* 41:17–18.

Ellis, E. 2011. "Neither Good nor Bad." *New York Times*, May 19.

Ghosh, A. 2016. *The Great Derangement: Climate Change and the Unthinkable*. Chicago: University of Chicago Press.

Hulme, M. 2014. *Can Science Fix Climate Change? A Case against Climate Engineering*. London: Polity.

Jacquet, J. 2013. "The Anthropocebo Effect." *Conservation Biology* 27 (5): 898–99.

Klein, N. 2014. *This Changes Everything: Capitalism vs. the Climate*. New York: Simon and Schuster.

Kolbert, E. 2011. "Enter the Anthropocene—Age of Man." *National Geographic*, March. http:// ngm.nationalgeographic.com/print/2011/03/age-of-man/kolbert-text.

———. 2014. *The Sixth Extinction: An Unnatural History*. New York: Henry Holt.

Leopold, A. 1935. "Why the Wilderness Society?" *The Living Wilderness* 1:6.

Lynas, Mark. 2011. *The God Species: Saving the Planet in the Age of Humans*. New York: National Geographic.

Malm, A., and A. Hornborg. 2014. "The Geology of Mankind? A Critique of the Anthropocene Narrative." *Anthropocene Review* 1 (1): 62–69.

Massey, D. 2005. *For Space*. London: Sage.

McClintock, A. 1995. *Imperial Leather: Race, Gender, and Sexuality in the Colonial Contest*. New York: Routledge.

Monbiot, G. 2010. "This Tory Bonfire of Regulations Lets the Rich Foul the Poor with Impunity." *The Guardian*, July 12. https://www.theguardian.com/commentisfree/2010/jul/12/ tory-bonfire-regulations-rich-foul-poor (accessed August 29, 2015).

Moore, K. 2013. "Anthropocene Is the Wrong Word." *Earth Island Journal* (Spring). http:// www.earthisland.org/journal/index.php/eij/article/anthropocene_is_the_wrong_word/ (accessed July 12, 2015).

Noah, T. 2013. *The Great Divergence: America's Growing Inequality Crisis and What We Can Do about It*. New York: Bloomsbury.

Nordhaus, T., and M. Shellenberger. 2009. *Break Through: From the Death of Environmentalism to the Politics of Possibility*. New York: Mariner.

OXFAM. 2013. "The Cost of Inequality: How Wealth and Income Extremes Hurt Us All." OXFAM.org, January 18. https://www.oxfam.org/sites/www.oxfam.org/files/cost-of -inequality-oxfam-mb180113.pdf (accessed November 22, 2015).

Plumwood, V. 1994. *Feminism and the Mastery of Nature*. London: Routledge.

Robin, L. 2008. "The Eco-humanities as Literature. A New Genre?" *Australian Literary Studies* 23 (3): 290–304.

Schuppli, S. 2014. "Material Witness." SusanSchuppli.com. http://susanschuppli.com/ exhibition/material-witness-2/ (accessed December 10, 2015).

Smaje, C. 2015. "Dark Thoughts on Ecomodernism." The Dark Mountain Project. http://dark -mountain.net/blog/dark-thoughts-on-ecomodernism-2/ (accessed December 21, 2015).

United Nations Refugee Agency. 2014. *World at War*. New York: United Nations.

Hubris

Anthropocene in a Jar

Tomas Matza and Nicole Heller

It was an ordinary afternoon in January, 2014. We were desperate to escape the monotony of North Carolina's thickly forested and flat Piedmont landscape. It was a place we had moved to only recently, a place that could feel claustrophobic. Without a particular goal in mind we set out by car, we two and our kids, for the expansive coast. We needed a vista.

We wound up at Wrightsville Beach—a two-and-a-half-hour straight shot down I-40 that passes through the Piedmont's strip malls and pine stands. It was a clear, sunny afternoon and it felt perfect to be on the long, empty edge of the Eastern Seaboard. Seagulls floated on the air currents, sanderlings scampered back and forth at the water's edge, following the washing waves in and out, in and out, their little legs blurring together into a grey smear. Although winter, it was warm enough to put on our bathing suits, and we began to play on the beach at the edge of the waves, digging into the sand.

What is it about digging holes in the sand? Something about feeling the hard granules under the nails? Or maybe the fantasy of coming up out of the ground in China?

Our digging immediately revealed curious swirls—layers of sand . . . shell . . . sand . . . shell. The fingerprint pattern was mesmerizing—and we found ourselves digging deeper and deeper, noticing the width of each layer, until the waves washed in, filled in the holes and we would begin again.

Our ever-curious eight-year-old son Aarno asked, "What causes the stripes?" And so we began to build an answer, collaboratively. Abstract earth processes turned into something palpable. "The stripes point to something cyclical," we said, "but what?—a tidal process, perhaps?" We imagined the moon's tug on the sea, and the wave's tug on the sand and shells, forming the layers, over time, gently. But how much time would a band of sand represent? We wondered . . .

Digging down, we admired the bits of shell. We separated big from small, pinks from blues. We tried to paste them together in order to reconstitute the animals—bivalves of various sorts, clams, oysters, mussels. We imagined the living ones hidden nearby, buried beyond our probing hands, in the marshy backwater lagoons, or farther still, out in the dark ocean.

Like so many children's questions (why is the sky blue?) we were unable to answer Aarno's with confidence, and we set out on a small research project. We knew that this area of coast had been the site of major beach engineering. Could that be the cause of the distinct layers? Or was this simply how beaches in this part of North Carolina had always been—the novelty of the stripes perhaps only a novelty to us newcomers? We found ourselves betwixt and between. Were the stripes natural, from waves and wind, or artificial, a result of beach engineering?

We would return months later to collect an object of wonder and concern—a sample of swirling sand and shell, captured in a recycled kimchi jar, now housed in a museum exhibit (plate 2). Eventually we came to understand that the jar contains a vast ecology of ocean cycles, tides and moons, wave dynamics, tunneling critters, barrier islands, lagoons, and debris from ancient mountains—things one could classify as "natural." And it contains pipes, dredging ships, dream houses, cars, carbon emissions, and people with their toes in the sand—things one could classify as "human." In essence, the recycled jar captures the entwined and often vexed processes of earth cycles and human development. This entwinement points to what scholars in the sciences and the humanities are calling "the Anthropocene," a concept signifying the increasing human-domination of earth processes. Our jar reminds us how difficult it has become to think of any earth process, whether oceanic, climatic, geomorphic, or otherwise, without also thinking of the human.

Is it disappointing to realize that little discoveries like ours—of the stripes— bear human fingerprints, and capitalist fingerprints at that? Or is it exciting to realize the complex ways that human and natural histories are entwined? In answering our son's question, we realized that it is not the entwinement of human and nonhuman itself that is disappointing (the two cannot be separated anyway), but rather the type of entwinement. The story of the stripes is our way into the Anthropocene—the Anthropocene in a jar.

* * *

A local real estate website lists the average price for oceanview or oceanfront housing at Wrightsville Beach at just under a million dollars. Averaging 1,600 square feet in size, this amounts to highly priced housing in this summer resort town. Moreover, for many months of the year more than half of the homes sit empty, and Wrightsville Beach—as on that sunny but cool January day when we dug in the sand—is mostly a ghost land (or a bird land, to be more precise).

But those few summer months are active. Thousands of people escape the humidity of the Piedmont and seek refuge at the coast. This seasonal migration of the beachcomber has been going on since at least 1905, when Wrightsville Beach was first developed as a resort "at the end of the line." And, as we learned, this seasonal pattern has depended on highly capital-intensive engineering. Since 1939 ships have dredged, pumped, tunneled, and dumped millions of cubic yards of sand and shell every four years to re-establish the rapidly eroding beach. Called "nourishment," perhaps much of the sand that got under our nails that January day was put there on twenty-five occasions to the tune of $4.3 billion. In the most recent $55-million nourishment, dredgers sucked up sand, shells, and anything else in their path from the nearby Masonboro Inlet and, using a three-foot-wide pipe that stretched two miles, dumped between 700,000 and 900,000 cubic yards of sand on the beach.

How much sand is that? Imagine a large tractor-trailer. Now line many of them up for a little over sixty-six miles. Connect them to a huge pipe and release their contents. It takes six weeks of constant spewing.

Rinse. Repeat every few years.

That is a lot of sand, and a lot of expense. Nourishment is also a major recurring disturbance to the local ecology. It can destroy, or temporarily displace, the features of this unique habitat that make it a safe and productive home for many creatures—from an endangered sea turtle to many of the colored bivalves that provide essential food for bird and fish populations and ultimately contribute to the remarkable stripes we discovered. Over many years of nourishment, many of these animals either diminish in abundance or are forced to seek refuge elsewhere

Wrightsville Beach's story is one found on beaches up and down the Eastern Seaboard. Sandy escapes from New England to Florida have been eroding for decades—a process that engineers have been trying to reverse for as long.[1] But beach nourishment is a Sisyphean task. Under undisturbed conditions, barrier islands like Wrightsville Beach shift back and forth, like a migrating organism. They dance flirtatiously alongside the coast, coming in close, and then pulling away. Sand taken from these islands on one day's oceanic binge will return as indigestion on another. Property owners call this "erosion," but its more precise name is "natural coastal dynamics."

In contrast to the current fixation on fixity, early settlements adapted to the ephemerality of the North Carolina barrier islands. Native American Siouan tribes, known by early colonists as the Cape Fear Indians, used these islands seasonally for fishing, hunting, and foraging. They built temporary camps that would wash away later. Euro-American settlers in the eighteenth and nineteenth centuries also lived with the barrier islands' shifting nature, tearing dwellings down and retreating inland after catastrophic wind- and sandstorms. These "bankers," as they were called, eked out an existence in an "isolated, cash-poor" and "volatile landscape" as described by historian Gabriel Lee (2008). Land use was informal and communal.

When the Wright brothers flew the first airplane at Kitty Hawk, North Carolina, a barrier island 170 miles north of Wrightsville Beach, they, too, were struck by the shifting sands:

> But the sand! The sand is the greatest thing in Kitty Hawk and soon will be the only thing. The site of our tent was formerly a fertile valley, cultivated by some ancient Kitty Hawker. Now only a few rotten limbs, the topmost branches of trees that then grew in this valley, protrude from the sand. The sea has washed and the wind blown millions and millions of loads of sand up in heaps along the coast, completely covering houses and forest. Mr. Tate is now tearing down the nearest house to our camp to save it from the sand.[2]

This view of the islands' shifting sands itself shifted in the early twentieth century. With the rise of beach resort culture and the big dollars it brought, the barrier islands were converted into fixed assets. As with other histories of enclosure, this patch of sand was turned into a form of property and developed into a playground for nearby North Carolinian urbanites. Its property form was anchored by an expansive material infrastructure: Roads, jetties, inlets, train tracks, contracts, beachfront hotels, and cottages were built to mold, secure and connect the island to the mainland, which previously had only been accessible by boat.

People don't like the sand under their feet—and especially under their beachfront property—to move. Once sufficient capital had been invested, the dance of sand became a source of government concern. Ironically, when the sands' travels were blocked by development, the barrier island's movements became even more pronounced. Thus the state began "nourishment"—a public works project largely paid for with tax dollars and enacted to protect private property. Beach nourishment pours billions of dollars into sand that sits for a while onshore, only to be retaken in an endless exchange between ocean and land.

* * *

Even a project as grand and disruptive as dumping hundreds of truckloads of sand two to three times a decade is unlikely to account for bands of shell separated by a few centimeters of uniform sand granules. On a return trip to Wrightsville Beach we saw that the striped pattern itself changes along the beach. In one spot our digging easily revealed perfect stripes, while in another they were missing. This spatial irregularity was also temporal: watching the lapping waves for a few minutes, we saw shells dumped on the surface, giving the beach a pimpled look, only to be swept away by another wave.

Eventually we found a wonderful book called *How to Read a North Carolina Beach* by geologists Pilkey, Rice, and Neal (2004). It is a beach lover's companion filled with all the whys, hows, and whats of North Carolina beaches. The answer to the mystery of the stripes turns out to be both much less, and much more, than what we thought. Indeed the stripes are "natural." The authors call them a "layer cake" produced by storms and fair weather. Basically, big waves (as in a storm) churn up the top layers of the beach with each pass. As the size and energy of the waves decrease, the coarse, heavy materials such as shells settle out first, followed by the finer sand. This simple process creates distinct layers. In a sense, then, the stripes are the signal—and a beautiful signal at that—that emerges from the noise of all the various processes involved.

Yet, even in their naturalness the stripes were also anthropogenic. Climate change has elevated the seas and the intensity of storms—two factors that affect the striped pattern's width and number. Beach nourishment has played a role, too. The composition of the beach—the sand and shell that make for the color, feel, and smell of place—reflects its geological history. Take color as one example. When clams, and snails, and other creatures die, the color of their shells changes depending on where they decompose. On the beach they eventually turn brown through the process of oxidation. But in a lagoon on the backside of a barrier island they turn black in the oxygen-poor mud of the marsh. Pilkey, Rice and Neal discuss how geologists have relied on the color and composition of the shells and sand to infer the natural history of the beach—how it was formed, and how it migrated and changed over thousands of years.

Dredging alters this pattern. Specifically, it takes substrate from different places and mixes it together on to the beach. Sometimes it takes sand and shell from lagoons—thus introducing lots of black shells that would otherwise not have been there. That means that beach nourishment effectively erases the natural history. We can't "read" the beach in the same way anymore. As a place, the beach—and our stripes—now reflect a modern history, a history of development and dredging.

Dredgers and pipes, bulldozers, carbon, and poorly chosen housing sites—together these point to a Rube Goldberg–esque assemblage. Human activity vies for attention in our puzzling along about sand on a beach. How do we feel about this? Is it disappointing to realize that little discoveries like ours—of the stripes—bear human fingerprints, that the beach is a human construct?

It is indeed disturbing to realize that the land under our feet is being changed in ways big and small by human activities, with little consideration of the impacts or, apparently, public expense. A turning away from other responsibilities underlies the public expenditure on beach nourishment. Recently the North Carolina state legislature cut funding for education, and the state's Department of Environment and Natural Resources forbade the use of sea level–rise projections in policymaking, at the same time that another publicly funded round of nourishment was approved.

The jar of sand thus indexes both violence and inequality. Beach nourishment works against, rather than with, earth processes, and its fleeting benefits are primarily for a wealthy few, while the broader effects of anthropogenic climate change on livelihoods are categorically denied. The welding of particularly violent human-nature relationships to social inequality has recently prompted historical geographer Jason Moore to argue that a better name for the "Anthropocene" might be the "Capitalocene." As he argues, "The Anthropocene makes for an easy story. Easy, because it does not challenge the naturalized inequalities, alienation, and violence inscribed in modernity's strategic relations of power, production, and nature. It is an easy story to tell because it does not ask us to think about these relations *at all*" (2013). As the story of Wrightsville Beach attests, it only takes a little digging to reveal the unequal social relations underlying beach nourishment.

And yet there is an opportunity to turn the jar of sand in a different direction: the very existence of the layer cake, now and in the past, reflects the constant, consistent process of wind and water shaping material in familiar ways. The engineers do not dump the sand and the shell in this way, yet the layers emerge nonetheless. The emergence of this pattern crystalizes the complex ways that human and nonhuman/cultural/geological histories are entwined and co-constitutive. As much as humans may try to diminish the earth's many cycles, they emerge and persist, like a weed growing through a crack on the sidewalk. Patterns like these have always been and will always be.

* * *

This jar of sand, for us, ultimately contains an ambivalence that lies at the heart of our current epoch. It is an ambivalence wrought as much from dreams as anxieties. The jar symbolizes particular humans, bound to a capitalist political economy, fighting for the stability of *capital* dreams, in the face of cyclic, repetitive, and chaotic earth processes. The fight sculpts the land into something novel—neither artificial nor natural but hybrid.

And the jar symbolizes folly—the false presumption that modern industrial societies, rather than deep time, will ultimately prevail, and that humans can rework the relations of earth and ocean according to the will of a few without consequence. In this state of anxiety, there must be a nagging intuition that this relation must change. And yet it feels impossible because so much has been invested in stabilizing and claiming land. The jar of sand whispers, "This land is mine, I bought it, and the state better secure my purchase."

Pointing through and also beyond this sense of ecological and theoretical impossibility feminist theorist Donna Haraway (2015) has suggested that we need to create refuges on Planet Earth for refugee species, turtles, bivalves and humans alike, who have been displaced by outcomes of expansive capitalism and violent engineering. These refuges almost by definition will have to be assembled from new relations between species, and new human-ecosystem process interactions. Like Haraway, we also think that something sustainable and diverse will only emerge through relations organized by collaboration and respect. Only through collaboration with the ocean will the beach shift from being a playground for the few to a refuge for the many. In a hopeful way the jar tells us that there are abundant instances for such interactions. We dump sand again and again, and each time other earth systems are ready to play—evolving, reconstituting, and shaping—in ways that are both familiar and new.

Overcoming the dystopic Anthropocene in this corner of the world means learning to live with mobile beaches. It means building temporary or storm-surge resilient bridges, houses, and roads, such as advised by the North Carolina Coastal Geology Cooperative Research Program in 2008. Going further, why not return barrier islands to a shifting commons—a human land-use form more in sync with the ebb-and-flow ocean cycles—rather than affixing them to the logic of private property? And why not take the public investment in beach nourishment and direct it to other, more pressing social or environmental needs, such as funding schools or combatting sea-level rise? It is the twentieth-century development of the North Carolina coast that is at odds with the fluidity of the barrier islands, not human beings per se. And it is a particular system of social and economic development that is at odds with the goals of environmental health and justice, not simply "*anthropos*."

NOTES

1   "Between 1922 and 2003, beginning with the first beach nourishment at Coney Island, New York, at least 970 projects have 'nourished' more than 6,050 kilometers (km) of US shoreline along the Atlantic and Gulf coasts, using 430 million cubic meters (m³) of fill" (Preston and Bishop 2005).

2   Orville Wright, October, 14, 1900, in McFarland 1953.

BIBLIOGRAPHY

Haraway, D. 2015. "Anthropocene, Capitalocene, Plantationocene, Chthulucene: Making Kin." *Environmental Humanities* 6:159–65.

Lee, G. F. 2008. "Constructing the Outer Banks: Land Use, Management, and Meaning in the Creation of an American Place." Master's thesis. Retrieved from http://repository.lib.ncsu .edu/ir/bitstream/1840.16/2509/1/etd.pdf.

Lerch, P. 2004. *Waccamaw Legacy: Contemporary Indians Fight for Survival*. Tuscaloosa: University of Alabama Press.

McFarland, M. W., ed. 1953. *The Papers of Wilbur and Orville Wright*. New York: McGraw-Hill.

Moore J. 2013. "Anthropocene or Capitalocene?" Jason W. Moore (blog), May 13. https:// jasonwmoore.wordpress.com/2013/05/13/anthropocene-or-capitalocene/ (accessed on September 18, 2015).

Pilkey, O. H., T. M. Rice, and W. J. Neal. 2004. *How to Read a North Carolina Beach: Bubble Holes, Barking Sands, and Rippled Runnels*. Chapel Hill: University of North Carolina Press.

Preston, C. H., and M. J. Bishop. 2005. "Assessing the Environmental Impacts of Beach Nourishment," *Bioscience* 55 (10): 887–96. http://bioscience.oxfordjournals.org/content/55/10/ 887.full.

Concretes Speak

A Play in One Act

*Rachel Harkness, Cristián Simonetti,
and Judith Winter*

An experimental script for a performance that reflects upon concrete's position within the epoch known as the Anthropocene. First performed in Madison, Wisconsin, on November 10, 2014.

CHARACTERS

CONCRETE CHORUS: A collective chorus, made up of multiple ambivalent voices. Its timbre is gravelly in places, smooth in others. It is heavy and largely low in pitch. With the paragraph breaks, different combinations of voices can come to the fore.

MAKERS #1 AND #2: While the chorus speaks, the makers prepare (gather) what is necessary for the operation and then carry out the mixing and pouring of concrete into a steel form (mold), followed by the early stages of its curing.

SETTING

A site associated with concrete construction or a public space. An assembly of people is envisioned in the audience, and it is to them that the CHORUS addresses itself. The MAKERS occupy the space while carrying out a choreographed mixing and pouring of a small concrete cube form. Their acknowledgment of the

CHORUS is implicit and comes across in the rhythm of their work and their quiet—as if to listen to the concrete's narrative.

The CHORUS may be played by onstage actor(s) or be represented only in voice(s) heard, for example via a recording played off-stage.

TIME

The present day.

PROPS*

one bag of cement (< 10 kg)

one bag of sand (at least twice the amount of cement)

one bag of gravel (at least three times the amount of cement)

one large bucket of water

one measuring jug (clear plastic, if possible)

one disassembled and pregreased standardized 150mm² steel "test-cube" mold (four
    sides, one base, four screws, four bolts)

one adjustable spanner

one hand shovel

one tamping rod

two small hand trowels (one gauging and one margin)

one cloth to clean with

one 2 × 2m tarpaulin

one 50 × 50cm piece of plastic sheeting

two standard work-wear boiler suits

two pairs of gloves, disposable masks and goggles**

* Props should be sourced from local suppliers and should reflect the common building culture's practices.

** Cement is caustic so MAKERS should decide on appropriate health and safety precautions.

Scene 1

(CONCRETE CHORUS is alone at first. The space, empty and dark.)

CONCRETE CHORUS

Our history is foundational to yours. Look out and there we are: from great dams and bridges down to simple pillars and posts. We are what you build upon, what you build with. Our development has enabled your growth.

We are concretes. The most used material after water in the world today: from Chile to Scotland, Malawi to China, our presence here on this planet is only increasing. The malleable strength of our mix of cement, sand, water, and aggregate has allowed us to rapidly consolidate our position as the most indispensable building material in modern construction. Paradoxically, this is perhaps because of our aggregate and varied nature. That is, we exist in all sorts of combinations of constituent ingredients and in all sorts of contexts of use. To hear us speak as one is therefore rather unusual—we are a congregation provoked or invoked by the arrival of this epoch of yours known as the Anthropocene.

(MAKERS enter the space, and put on their boiler suits and other safety equipment [optional] as the light on them increases. They bring the props to the center-front of the performance space and begin the process of making a concrete cube. Lighting focuses here. Regarding the pacing of their activity, these directions only indicate what stage the process may be at a certain point in the chorus's narrative.

They lay down the tarpaulin first and then carefully set up their tools and materials in an array in front of themselves and the audience. They move quietly and calmly, but with purpose.)

CONCRETE CHORUS (*CONTINUED*)

In an earlier era, the Greeks and Romans discovered the power of a volcanic rock known as the Pozzolanas and, from this siliceous and aluminous mineral, evolved us for use in buildings and underwater construction. In time, our human makers found ways of creating us without need for volcanic ashes and the process of making modern Portland cement was invented. This process uses very high temperatures to convert finely ground materials such as limestone, clay, and shale into a mixture containing the four key ingredients of cement (calcium oxide, aluminum oxide, silicon dioxide, and iron) as well as a byproduct of carbon dioxide. Then processed in a long horizontal furnace called a rotary kiln, the ingredients undergo chemical changes to form a glass-like material called clinker. Gypsum (calcium sulfate) is then added to the clinker and they are ground to a fine powder: cement.

Cement is that crucial "glue" of our concrete mix, but it's the addition of water that activates its chemistry! Water hydrates cement's calcium compounds to form new compounds that bind the gravel or crushed stone "aggregates" into concrete.

(One MAKER takes one part of cement, two of sand, and three of gravel from the bags, using the shovel, and places them in a mound in the middle of the tarpaulin. S/he then mixes them by hand. The MAKERS focus on the handling of the materials and equipment. Meanwhile, the second MAKER assembles the test-cube mold with the spanner.)

CONCRETE CHORUS (*CONTINUED*)

Considering cement production for a moment longer may help us press upon you what a fundamental role we've played in this geological period marking your species' impact on the planet. Via cement production alone, we already contribute between 5 and 10 percent of anthropogenic carbon-dioxide emissions. Carbon dioxide may be invisible to the naked eye, but we add more to the earth's atmosphere than global air transport does. In fact, with one ton of cement per person per year being produced, each ton is releasing up to a ton of the carbon dioxide that is warming this planet. Current consumption rates of concrete are only set to rise, perhaps doubling in the next thirty years. Despite this, the stories of our production, use, and impact are not as widely recounted as they could be, considering our many entanglements with you. We are, in fact, a relatively unexplored icon of this era. If the Anthropocene is a register of humanity's reshaping of the earth, then we are a ubiquitous, almost omnipresent manifestation of this phenomenon.

In this somewhat unusual address, we have claimed a voice and are recounting our process of becoming, our history, our relationship with you, and our role in this, your anthropogenic, anthropocenic era. Why should you have the only say here, we thought! The silencing of things and materials, and, more widely, of the more-than-human elements of this world, has arguably contributed to the current environmental crisis—a crisis driven by humanity's unlimited desire to shape and control nature. The event of our speaking may open eyes to our ubiquity and our impact, or at the very least communicate a sense of our ongoing and formative presence in the world we share with you, long after the rumble of the cement mixer has died down.

(The MAKERS place the assembled cube to one side in order to make room for the mixing. One MAKER adds the water, judiciously pouring from the jug, as the other draws the dry materials in and begins to mix.)

CONCRETE CHORUS (*CONTINUED*)

In comparison to you nimble and dynamic creatures, we concretes seem relatively silent and slow-moving. We also seem homogenous: "Oh, it's concrete," you say, referring to one of us as if it is much like every other. But we are many and varied! We're determined by our specific maker's imagination and workmanship, by our situation and its climate, by what's required of us, and by our constituent materials—which can vary hugely. We are the complex process of concretizing and the product of much preparation, labor, and craft. We are also something lively, moving and changing over long timespans. We're capable of forming things that are at once extraordinary and prosaic.

(When enough water is added, both MAKERS start to mix the concrete for some time: turning it, scraping and lifting, pushing the mix so that all of the materials are integrated. At this moment the odor of the concrete is present in the space, as well as the sound of the mixing of the aggregates with the cement and water.)

CONCRETE CHORUS (*CONTINUED*)

Our composite nature means that we are most often mixes of fine grey cement powder, precious water, and the granular components of sand and stone (carefully sieved and graded for relative coarseness or fineness). Amounts of these ingredients are mixed together in delicate balance, often having traveled great distances before their meeting. You can think of us concretes in terms of concrescence, then, or a growing together. As much as we are distinct things in the world, or as much as we become them, we are also gatherings. Then, under stress and rain and use and neglect, time wears us down or changes us up! Concrete as gatherings—although often used in order to resist the passage of time—are therefore impermanent, always moving and changing.

(MAKERS concentrate on the physicality of the action of mixing the concrete together for a few minutes as they gather a rhythm that works for them and with the chorus's voice. The mixing produces a considerable rasping-churning noise.)

CONCRETE CHORUS (*CONTINUED*)

In fact, we spend a good part of our existence in a molten state—reminiscent of our volcanic origins (plate 3). Here, we are kept liquid and moving, often turning endlessly in a barrel, round and round. Once poured into a mold (which we call a form) we can then become solid at varying rates. To say that we are *solidifying* might be a more accurate way of putting it, actually, because our state of *not-yet* solid lasts a considerable time. At this moment we require care. Our appearance might look solid, but we ask to be kept liquid still—gently misted with water or covered under blankets of polythene. The longer you allow us to remain in this formative state, the stronger we will grow.

So although it is often solidity that is used to characterize us, it's not as simple as this. Since ancient times humans have been calling us *liquid rock* or *liquid stone*, a name that suggests an uncanny, magical substance. The fact that your ancestors named us with such an oxymoron suggests that although you build your viaducts and monuments as if they would last forever, you know, somehow, that while we are solid we are also *not*.

Surely enough, at this point you may be thinking, "Won't this very performance before me produce a solidified thing?"

Yes, and no. This performance will leave just the trace of the process of concrete's production, and while that trace will consist of a concrete thing (a cube in fact) our intention is not to present ourselves as a solid static cast object. Instead, it is to offer the onlooker a chance to see us from a different perspective: to see us in our various states, molten included. We wish to recast the history of modernity! Here we make visible the process of our mixing, pouring, and curing, and, particularly by presenting the experimenter-concrete-makers, we seek to draw attention to the craft and labor essential to our production. We may have a voice here for a time, but we are interdependent with you, our makers.

(A MAKER places the assembled form [test cube] back at the center of the tarpaulin once more, next to the concrete. Both MAKERS consolidate the mix, pushing it into a pile as they prepare to pour it into the cube mold.)

CONCRETE CHORUS (*CONTINUED*)

To return to this cube for a moment (to this modest yet sharp-edged and weighty thing which will be turned out from the stainless steel form you now see bolted together and ready to be filled): this "exhibit piece" will be our representative—one form and version of concrete standing in for many more—a tiny component that might testify to the massive scale of concrete use.

The 150mm^2 cube of concrete that this form molds is known as a "test cube." It is arguably a "test" cube in several senses of the term. Firstly, the cube mold is a ready-made form used worldwide in the construction industry—you'll as likely find these in Mumbai as in Munich. They are used precisely to test whether builders have their mix of ingredients just right: each cube is tested for its compressive strength in a controlled and standardized way. So, unlike the finished blocks that builders might buy "off the shelf," so-to-speak, our representative cube awaits its own destruction in this quality testing. Eventually it will be wasted, broken to rubble, and will contribute to one of the many mountains of demolition wastes amassing around the world.

In a second and third sense, this is a "test" cube in that its production is experimental play rather than an example of seasoned practice, and the cube-making is a sampling of a vast and diverse world of everyday human concrete practices and productions. Finally, the "test cube" is experimental in the sense that getting the right mix for us concretes is crucial. If our proportions are wrong, or we are inadequately mixed or neglected while curing, then we will not hold together. Our every mix, pour and cure is part of a long history of trial and error.

(MAKERS pour the mix into the test-cube mold. One removes the excess, then the other tamps the mixture once it is in the mold to release air bubbles trapped within it. This is done for a minute or so. These actions create their own rhythm and sound within the space.)

CONCRETE CHORUS (*CONTINUED*)

Speaking of this history, we concretes have existed for a very long time. As far back as 6500 BC the very earliest human-made concretes—for floors—were mixed from lime, sand, gravel, and water. There have been periods of history where we have fallen from use and you have all but forgotten us. Then there have been glorious rediscoveries and burgeoning applications that have redeemed our popularity. For instance, the Moderns dreaming in concrete from the 1920s onwards revealed what we concretes could afford, pushed us to our limits, celebrated us in their monumental constructions. Today, mirroring a little of this spirit, we are still part of an open-ended process of innovation and experimentation, where new formulas are continuously being developed, tested, and adopted. The aim now is to reduce our carbon emissions.

For instance, some recent evolutions have been into mixes that will produce transparent, self-healing, and "green" concretes (including those made from hemp), even carbon dioxide–absorbing concretes. Whether or not these new forms become widely adopted will likely depend upon how flexible humanity's habits are and what economic values you hold to. In our experience, the development of these newer forms has depended upon the monetary cost involved in the process, the potential for the creation of profit, and the convenience of the use of these new forms. So, although we exist as a reservoir of possibilities in these new versions, at the moment our potential remains nascent, really only part of the imagined futures of concretes.

Those futures (ours and, by implication, yours) are likely to be shaped in large part by geopolitics, by cultures operating at worldwide scales. These forces determine our production and use across the globe. The amounts in which concretes are produced and consumed are in almost perfect correlation with what your World Bank describes as "economic development." Thus, Europe and North America have historically dominated our industry, while the entire continent of Africa, for example, possessing the smallest economy of the three, is also the smallest consumer and producer of concrete. In recent years the fast-growing economies of China and India have seen their concrete use rise meteorically: we now move globally.

(Once the tamping is complete and any excess is scraped from the cube mold, one MAKER lifts and drops the corners of the cube a little, one at a time—with resulting thuds. S/he does this several times to release any air trapped in the mixture.)

CONCRETE CHORUS (*CONTINUED*)

Because the prices of cement are controlled by powerful international cartels, local building economies are vulnerable and dependent upon the movement of global economies to ensure the availability of our constituent materials. Incidentally, this is a long way from how we used to be made. Previously we would have been crafted from local materials: from clinker fired in smaller-scale lime kilns and aggregates determined by what could be collected from nearby sources. Local building cultures, too, have over time mostly now abandoned other home-grown methods and materials of construction in favor of concretes in large-scale production. The extent of clamor and pressure for the "modern" that is still heavily associated with our use should not be underestimated.

Attempts to change our constituent materials for green ones, or—even more radically—to use *fewer* concretes therefore threatens to jeopardize the political-economic status quo. It threatens to impact upon humanity's self-imposed rush to develop or "improve" living standards around the world and to "modernize" through particular forms of construction. Clearly, we concretes do not mean the same thing everywhere in the world: concrete foundations that can survive earthquakes differ from waterproofing cement plasters; skyscraper foundations in financial districts differ from defensive military structures. It could even be argued that as we have become available to wider publics and easier to use, concretes have democratized architecture. With steel and glass we have certainly helped to revolutionize it. However, what remains—despite the differences—is the fact that we are a powerful force in the transformation of the world over time. The labor involved in our use determines the working days of millions of people whose bodies toil with us. The economies that we create, support, and rely upon shape your societies. The gases we release in our production, the extraction of our raw materials, the energy required to standardize us and transport us here and there, the landscapes we form when poured out over the earth: in all these ways, and more, you shape us and we shape you and the planet.

What also differs is the impact that we have upon the environment and the responsibility people take for this. In fact, we are increasingly at the center of human legal and moral debates concerning climate change and justice, and we hear of schemes called carbon credits, and battle lines being drawn somehow around the equator.

(MAKERS concentrate for a few minutes on carefully smoothing the upper surface of the packed mix inside the cube, using their trowels.)

CONCRETE CHORUS (*CONTINUED*)

We concretes epitomize modernity, then, and all that modernity has allowed humanity to believe, to build, and to bury. In this epoch known as the Anthropocene, our ubiquity coupled with our very contradictory characteristics as liquid rock mean that we seem both natural and artificial. Our ancient volcanic origins and the crafting of our mixture from local sands and stones connected us with nature. However, the manufacture of our cement seems to somehow catapult us forward in time, above nature, making us part of that paradoxical and precarious state you have built for yourselves. Using us, you have attempted to defy and control nature through massive constructions—such as skyscrapers that tower above, or sea walls that shelter and defend. You use us to mark and maintain boundaries: to contain, control, and divide human populations. In our infrastructural roles, you depend on us to enable a globally connected human world: harbors and jetties, communications towers, transatlantic channels, bridges and roads, all offering durable, smooth, hard, and flat surfaces on which vehicles of all sorts quickly move and take off. Concretes have allowed the constant acceleration of modern lifestyles.

But our discovery and use throughout the ages has facilitated even more promethean accelerations. With their Portland cement inventions, those nineteenth-century Victorians must have felt as if they had somehow discovered how to manipulate the earth, change geological history, without having to travel to the underworld to beg favors from the old gods. In a way, we concretes allowed for deep time to be compressed. And yet, we never could—and never did—promise to last as "natural" rocks do. Our relatively quick creation and our composition can hardly be compared to that of the rocks forged through the planet's long-term cycles of melting, erosion, and compression.

You seem to think us impenetrable, and because of this or perhaps because you have believed in the magic of liquid stone a little too fervently, you've built in concrete as if the future was yours for the taking, as if our solidity was homogenous and assured and could provide secure foundation for ever more daring and ambitious projects at larger and larger scales. Perversely, one consequence of this is that you have neglected your concrete creations somewhat: many of us now lie mossy and cracked, our skeletons of steel rusted, our façades graffiti-adorned. Perhaps you were imagining that we were just like our stony predecessors that you used to quarry and assemble by the block? Surely now you know that we depend on continuous relations of care, extended over time?

(MAKERS carefully and methodically tidy the scene to remove evidence of the process, removing all props from the stage apart from the cube and the plastic sheet. They clean any spills on the outer sides of the mold [test cube] and move the test cube with its concrete fill up from the floor onto a plinth. Then they retreat from the stage to leave the cube spotlighted and the chorus speaking.)

CONCRETE CHORUS (CONTINUED)

Even with care, no amount of curatorial practice, no amount of innovation, can set time still for us. The concrete constructions that were raised in times of prosperity and with a confidence that could be mistaken for arrogance are now falling. The halcyon days of early Modernism now seem full of misplaced optimism or perhaps naïveté. Now both the Modern and we concretes have such heavy connotations, now that we've been solidified and thus weakened in the imagination of men.

These ruined concretes (our decrepit selves) that increasingly serve as adventure playgrounds for the explorer-photographer speak of the failure of socioeconomic systems and a lack of understanding of the qualities of strength and longevity as much as they do concrete's own material weaknesses. The modern ruins that increasingly surround you—that are disintegrating, sometimes slowly, sometimes dramatically, back into aggregate—these ruins are testament to the fact that we can't live in the eternal present you envisioned for us and so for yourselves. We exist in the human timescale: components gathered, mixes mixed, and slabs poured by individual humans and specific machines. But we also exist in the geological timescale, and particularly within it we are revealed as liquid.

From rubble we came, and to rubble we will return!

(*Blackout*)

NOTE

The research behind this script has been supported by the projects "Knowing from the Inside" (funded by the European Research Council), "Solid Fluids in the Anthropocene" (funded by the British Academy for the Humanities and the Social Sciences), and "Concrete Futures" (funded by Fondo Nacional de Desarrollo Científico y Tecnológico, Chile, Nº 11150278).

BIBLIOGRAPHY

Bennett, J. 2010. *Vibrant Matter: A Political Ecology of Things*. Durham, NC: Duke University Press.

Davis, H. 2006. *The Culture of Building*. Oxford: Oxford University Press.

Forty, A. 2012. *Concrete and Culture: A Material History*. London: Reaktion Books.

Harkness, R., C. Simonetti, and J. Winter. 2015. "Liquid Rock: Gathering, Flattening, Curing." *Parallax* 76:309–26.

Harvey, P. 2010. "Cementing Relations. The Materiality of Roads and Public Spaces in Provincial Peru." *Social Analysis* 54 (2): 28–46.

Heidegger, M. 2001 [1971]. *Poetry, Language, Thought*. Translated by Albert Hofstadter. New York: Harper Perennial Classics.

The Age of (a) Man

Joseph Masco

If there were an emblematic text for the Anthropocene, a cultural statement so singular as to eliminate any doubt about mid-twentieth-century ambitions for the industrial transformation of the natural world, it might well be the 1973 Atomic Energy Commission (AEC) film *Plowshare*. A survey of the previous fifteen years of work on "geographical engineering," the twenty-eight-minute film offers a comprehensive portrait of the fusion of nuclear weapons and industrial capitalism. It conjures a world of white masculinity armed with an engineering mindset and a planetary vision (plate 15). The earth surface is presented as an inherently faulty design, one that can now be corrected in the name of both convenience and profit. Or as the invisible, godlike narrator of *Plowshare* intones in that confident mid-twentieth-century baritone:

> To bring water and food where there is only parched earth and people where there is desolation; to bring freedom of movement where there are imposing barriers and commerce where nature has decreed there will be isolation; to bring forth a wealth of material where there are vast untapped resources and a wealth of knowledge where there is uncertainty; to perform a multitude of peaceful tasks for the betterment of mankind—man is exploring a source of enormous potentially useful energy: the nuclear explosion. He sees the potentials and he sees the problems. To investigate both, and to develop the technologies that will turn potentials into realities, the United States is conducting for the benefit of all nations a program it calls Plowshare.

The film documents the engineering attitude of the mid-twentieth century, at the very start of what is now known as the "Great Acceleration" of human consumption, that exponential expansion in travel, industrial production, and globalization that generates a compounding problem for earth systems today. The invention of new synthetic chemicals and the extensive use of plastics were key parts of an emergent, petrochemical-based American society. Each innovation in this chain of technological relations not only enabled a middle-class-consumer lifestyle, it also contributed in some way to the carbon emissions that have subsequently shifted the chemical composition of the atmosphere with rebounding effects across oceans, icecaps, and climate. We cannot understand our current planetary condition without interrogating this historical fusion of revolutionary industrialism and nuclear-powered American nationalism.

Plowshare offers a future-perfect version of nuclear science, conjuring the possibility that nuclear explosions could be made "safe," that radioactive fallout could be eliminated, and that the serious work of remaking the geology of the earth would be merely the next act in an unfolding technological revolution involving better health, energy, and economy (see Kirsch 2005 and Kaufman 2012). The conditional persists through the film—"may," "perhaps," "could," "might"—these are factual qualifiers, but they are overwhelmed by the narrator's enthusiasm for remaking the surface of the earth and thereby enabling a new kind of species supremacy based in expert knowledge:

> Before each experiment, experts in geology, seismology, hydrology, meteorology, radiobiology, and many other fields bring their specialized knowledge and equipment into the field. Working with public health authorities they assure that the specific experiment is being conducted within accepted safety standards. This same thorough application will precede the actual applications of nuclear explosions wherever and whenever they may be. What are these actual applications? Some will be dramatic in their effect, as nuclear explosions move huge masses of earth in excavation jobs, reshaping the geography of the land in dimensions never before possible, to meet the needs of man, needs he can see as he struggles against the geography nature has pitted against him.

Plowshare offers an emancipatory narrative for a world not perfectly arranged for global capital, a technical means of resolving unruly natural formations through the combined work of earth scientists and nuclear-weapons experts. The program ultimately conducted thirty-five nuclear detonations between 1961 and 1973, in addition to a wide range of chemical explosive tests at many sites. These experiments sought to demonstrate the industrial potential of "underground engineering" for extractive industries or "excavation applications" for improved transportation. Geographically, the tests ranged from Mississippi to New Mexico to the Rocky Mountain states to Alaska, with a majority taking place at the Nevada Test Site. Improving nature—narrated as a conquest—is cast in *Plowshare* as a "peaceful" project, a progressive effort to convert a weapon of mass destruction into simply an everyday tool. The atomic bomb is used to mobilize the next stage in a colonial-settler project, one that positions the western frontier as a geological problem, as well as an incomplete territorial pursuit.

If the Plowshare program existed for fifteen years in the conditional mode, it nonetheless made nature a theater for active nuclear science. The detonations were in the world, producing environmental destruction and multiple forms of fallout (Masco 2015), contributing to the transformation of the global biosphere in the nuclear age. *Plowshare*, despite its stated commitment to safety and precision, performs turning such unwanted effects into externalities that, again in the future perfect conditional, would ultimately be eliminated via the next stage of research. Thus, the world is rendered as malleable and nuclear science is rendered as perfectible in the same anticipatory gesture. The future as imagined by Plowshare proponents, however, was not the future that emerged, as industrial effects accumulated across the biosphere to produce not an ever more secure world economy but a planetary distortion in earth systems visible today in rising temperatures, melting ice caps, an acidifying ocean and species extinctions.

Plowshare was a promotional film aimed at congressional funders; it sought to recruit viewers to a world that did not yet exist but that seemed to be already in sight. The film conjures up a vision of a nuclear-mediated economy, but one not ruled by the nuclear fear that dominated the Cold War. Listen to the promissory note offered to energy companies about the benefits of geographical engineering through nuclear detonations:

> Nuclear explosions deep underground break and splinter huge areas of rock. These massive effects may permit highly promising recovery of resources that have been impossible, or economically impractical, to extract from the earth. . . . They can see how nuclear explosions could increase the total recovery and rate of recovery of vast natural gas and oil reserves by effectively breaking the rock so that these valuable resources can flow through. They can see how copper could be extracted from the ground more efficiently, as nuclear explosions shatter the surrounding rock, letting through the solutions that dissolve the copper and carry it to the surface. They can see how the huge area of broken and fractured rock can be used for receiving as well

as releasing materials, for storing natural gas near its market areas, for storing rain-water that could seep down underground where it would not evaporate, or even for storing chemical waste materials in underground formations.

In the end, this kind of nuclear fracking did not prove profitable but the concept endured. Some of the experts on Plowshare continued to pursue technological breakthroughs in nonnuclear hydraulic fracturing, ultimately enabling the shale formations from Texas to North Dakota to the Canadian Tar Sands to be developed by energy companies in the early twenty-first century. Indeed, fracking of the kind proposed in *Plowshare* is currently projected to make the United States the world's leading energy producer. Thus, even in failure the Plowshare project contributes to a future of expansive energy exploration and extraction. These innovations, and decades of petrochemical-based consumption and industry, now present a profound challenge as they have destabilized a planetary-scale climate system.

The kind of profit/loss calculation depicted in *Plowshare* assumes a universal value system, an embedding in petro-capitalism, and a vision of undeveloped land as waste. This notion of a singular human economy was challenged by local populations wherever Plowshare experiments were proposed. The first major effort, Project Chariot, was an attempt to build a new channel at Cape Thomson in Alaska. The idea was to detonate five nuclear devices in a row and create both a new harbor and channel connecting the harbor to the ocean. Extensive studies were made of the geology but local communities, including Inuit populations with cultural ties to the area, were not consulted. The project conducted multiple tests with high explosives, creating lasting environmental problems but the nuclear component was ultimately defeated by an unusual coalition of indigenous groups, activists, and biologists who challenged the Plowshare safety plan at each step (see O'Neill 2007). Project Chariot was halted in 1962. This vital history of environmental activism against Plowshare does not appear in the 1973 film despite its comprehensive overview of the program. This silence suggests that public protest could be cast by the AEC as another kind of externality to geographical engineering, one more thing to be overcome in the perfection of the science. The failure of Project Chariot was largely due to public concerns about radioactive fallout. By the 1960s activists could draw on an extensive public record of environmental damage from nuclear testing in the US Southwest and Marshall Islands dating back a full generation. And indeed, a number of Plowshare tests produced unexpected outcomes—from elevated fallout levels to local resistance—making the experimental program a key node in the development of a global antinuclear, peace, and environmental justice movement.

As the likelihood of finding a space within American territories for a large-scale demonstration of geographical engineering diminished after the demise of Project Chariot, Plowshare administrators looked internationally:

> But there is no doubt that most applications of nuclear excavation would be, not in the United States, but in other countries. The most dramatic example so far is in Central America: the blasting of a sea-level, Atlantic-Pacific interoceanic canal. Studies are being planned for both conventional and nuclear excavations on four possible routes for such a cut across Central America to supplement, and eventually replace, the Panama Canal, where ships now wait long hours to strain through the narrow, complex lock system, and others can't make it through at all. Before long it will be inadequate. Even before it was built half a century ago the complexity and limitations of this lock-type canal were realized. Men dreamed of a sea-level canal but it remained a dream. Plowshare may be able to make that dream a reality. And it is being considered. It is estimated that for certain routes, nuclear explosives could excavate the sea-level canal at one-third the cost of conventional excavation and in considerably less time. The end result would be a much wider and much deeper channel. A nuclear-blasted route across Central America could produce a navigable channel 1000 feet wide and up to 200 feet deep at mid-channel, offering a virtually unlimited capacity. No wonder this enormous project has so stimulated the imagination of the world, for a canal of this immensity, representing years of planning and development, complex engineering, and precise execution would be one of the greatest civil engineering feats of all time.

The Panama project generated immediate local fears of environmental damage, an index of the growing international antinuclear movement, which linked indigenous groups, citizens, and experts from within the broader earth scientist community (see Lindsay-Poland 2003). This notion of "empty" land ready for commercial development was thwarted by diverse interests who were already imagining a global ecology in need of protection, and by those with longstanding cultural investments in maintaining nonindustrialized territories. Plowshare, in other words, which had been a zone of pure engineering potential and geopolitical abstraction, became immediately politicized whenever it touched down in a specific place, fomenting environmental movements that continue to this day. The AEC argument that geographical engineering through nuclear explosives was faster and cheaper than any other industrial means was met by those with increasing experience of nuclear effects in a global biosphere transformed by Cold War military science, and who fought for different understandings of both value and profit.

* * *

Plowshare as a historical document reveals, however, the planetary spectacle of the nuclear age itself. It presents the rush for monumental projects and the desire for ever greater evidence of human ability to transform the natural world. *Plowshare* may well be unsurpassed as a cultural expression of hubris but it also marks an early commitment to what we would now call geoengineering—the effort to consciously transform earth systems to make them more amenable to human needs. Lawrence Livermore National Laboratory remains a pioneer of geoengineering research, linking Plowshare to a vast number of contemporary proposals to cool the planet through direct intervention into the atmosphere (see Hamilton 2013). Thus, the collective environmental future—with all its radical contingency—continues to produce programs for mechanical correction and even fantasies of planetary-scale control.

Plowshare thus has a once and future potential, offering a slippery challenge to our historical moment. Indeed, perhaps that is its essence: *Plowshare* we should remember was produced on 16mm film stock, a petrochemical medium that is both a form of fossilized time and an outmoded technology in the twenty-first century. The emulsions that allow filmic vision derive from the deep geological history of the planet. The dead plant and animal life that accumulated in watery sediments and was subject to vast pressures and tectonic shifts in the geological periods we now call the Carboniferous, Permian, Triassic, and Jurassic become the essential chemicals for analog photography. Thus, a motion-picture film about monumental earthmoving offers an enfolding set of temporal registers as the deep geology of the planet produces the very materials that enable the cinematic remaking of mountains and rivers as well as the promise of even more petrochemical discovery in shale. Watching *Plowshare* today offers viewers an opportunity to visit an outmoded future of geoengineering that may also be a portent of what is to come. The film's stylistic anachronism begs the question: what is actually different today in the realm of extractive industries, other than the now odd language of "geographical engineering"? What, if anything, has changed in human ambitions for geological extraction and planetary control?

So here we might pause for a moment to ask: what was that Enlightenment dream again—the one about human mastery of nature, accelerating revolutions in science and technology, and the ultimate perfectibility of Man? In the mid-twentieth century, the splitting of the atom seemed to supercharge this imaginary in the United States, signaling the imminent arrival of a superabundance, promising continuing breakthroughs in health, energy, and a consumer economy. After 1945 Americans constituted a dream world that, if it did not end in the fiery flash of nuclear war, would push relentlessly and inevitably toward a perfected capitalist society. This was the first "age of man"—a nuclear-powered fantasy that miraculously transformed an unprecedented destructive force into the expectation of a world without limits. The new rational order of science and engineering would remake everyday life in all its qualities, generating a series of new frontiers to be sequentially colonized, linking extractive industries to new urban landscapes to global communication and transportation systems to outer space in a spasm of industrial-scale world-making. Pause, just for a moment, to consider the intoxicating rush of this enterprise, the creative energy of making things that work on this kind of scale, of believing that people could finally shape reality rather than merely submit to it. The nuclear revolution promised to remake war, health, energy, and security, inaugurating a new golden age of human achievement in which longstanding social problems would fall in rapid order to the combined achievements of American technoscience.

The physicist Edward Teller, chief architect and advocate of the Plowshare program, put it succinctly to a broad American readership in *Popular Mechanics* magazine: "We're going to work miracles" (1960, 97). For Teller, weather control is not out of the question in 1960, and the global environment is positioned as an unruly domain that will be sorted out in short order via the explosive power of nuclear science. His vision is galactic and transcendent, identifying an earthly landscape to be remade for human commerce and convenience, as well as a Communist foe to be endlessly fought via military nuclear power. As perhaps the most vigorous twentieth-century advocate for nuclear weapons, Teller pushed for hydrogen-bomb development in Los Alamos before the first atomic explosion was achieved in July 1945. The United States built Teller his own Californian research facility in 1952: Lawrence Livermore Scientific Laboratory was designed to be competition for Los Alamos in the field of nuclear weapons science, but also as a place for Teller to pursue his vision of the nuclear revolution.

For the next forty years, Teller argued in favor of one high-tech weapons system after another, and against every nonproliferation and disarmament effort. Indeed, he promised presidents from Truman to Reagan that weapons science would fix the problem of the nuclear age—the minute-to-minute possibility of global nuclear war—not by eliminating the bomb but by perfecting it. Resistant to test bans and dismissing radioactive fallout concerns, Teller provided a rationale for a vigorous arms race with the Soviet Union while also promising a world remade via the benefits of nuclear science. By the 1980s he convinced President Reagan that the way out of the Cold War nuclear danger was to embrace yet another technological revolution, a third wave of nuclear science after those of the atomic and thermonuclear breakthroughs (Broad 1992). Teller proposed to surround the earth with space-based lasers that could destroy Soviet missiles on launch in midair, a proposal that President Reagan ultimately called the Strategic Defense Initiative (which in lesser forms continues as a research project to this day without objective success despite hundreds of billions of dollars in research). Reagan ultimately chose to pursue "Star Wars" over President Gorbachev's proposal that the United States and USSR simply dismantle their nuclear arsenals by the year 2000 (FitzGerald 2000).

In his commitment to remaking the world through nuclear weapons science, Teller approached the earth system as something that was filled with vital resources but deficient in significant ways. In 1958 he first proposed the concept of "geographical engineering" to overcome such natural barriers to commerce, launching Operation Plowshare at Lawrence Livermore Scientific Laboratory (Teller et al. 1968). The timing of his proposal was sublime, coming in the midst of worldwide fears about nuclear war as well as amplifying worries about the health and environmental effects of radioactive fallout from atmospheric nuclear detonations, and as the United States and USSR began discussions about a test-ban treaty. Thus, Plowshare offered a positive image of nuclear detonations in the midst of global nuclear fear, offering a vision of a world that—if it could just avoid nuclear war or permanently damaging the biosphere and humanity via radioactive fallout—might be a utopia for industry, commerce, and society. At a historical moment of maximum nuclear terror (with biologists discussing the effects of nuclear testing on the human genome, civil defense asking citizens to regularly rehearse nuclear war, and a geopolitics of nuclear intimidation underpinning Cold War conflicts around the world), Plowshare was offered as a means of recapturing the utopian potential of the bomb. Teller sought to publicly transform the bomb from a global menace to a vital form of capitalist creative destruction.

In the current expert search for the golden spike of the Anthropocene—that indelible marker in the strata of planet Earth that will end the epoch of Holocene in favor of a new Age of Man—earth scientists are increasingly focused on the radioactive signature of early Cold War nuclear explosions. Atmospheric detonations distributed cesium, strontium, and plutonium throughout the biosphere, leaving a marker of twentieth-century nuclear nationalism throughout the biosphere and in every earth system. Earth scientists report that plutonium from atmospheric nuclear explosions will be "identifiable in sediments and ice for the next 100,000 years" and have identified a sharp spike in cesium and strontium and plutonium beginning in 1952 with the first thermonuclear detonation (see Waters et al. 2016). This Anthropocene—this new Age of Man—by infusing political time with geological time now threatens to swallow whole historical eras like the Cold War, transforming that planetary-scale national competition into a permanent post-natural formation.

November 1, 1952, is the date on which the first thermonuclear explosion, known as Ivy-Mike, was detonated by the United States at Eniwetok atoll. Designed by Edward Teller and Stanislaw Ulam, it produced a ten-megaton detonation that created a mushroom cloud 25 miles high and 100 miles wide. The fallout from Ivy-Mike circled the globe and remains so comprehensive it can serve today as the key illustration of planetary-scale industrial effects. This explosion is now part of a nested series of temporalities: it is the start of a thermonuclear age inside an already established atomic age, a key moment in the Cold War, now poised to be the anchor for a new geological epoch, the Anthropocene.

Teller's vision of a world transformed by nuclear science has come true but in a highly perverse fashion. For we all now live in a world that is still capable of nuclear war and that is marked by the plutonium, strontium, and cesium of Cold War–era nuclear weapons tests. It is a world still committed to ever more dangerous forms of resource extraction (deep-water drilling and hydrological fracturing) and is increasingly interested in a geoengineering fix to the resulting damage to the earth system from a petrochemical-based economy. Transformed by the nuclear modernist visions of the mid-twentieth century, earth systems are now influenced in ways subtle and profound by industrial activity. Thus, perhaps our increasingly dangerous era is best thought of as a specific industrial-modernist achievement—the "Age of Man" as the materialized dreamscape of one radical but highly influential man—we might call it: the Teller-ocene.

BIBLIOGRAPHY

Atomic Energy Commission. 1973. *Plowshare* (film). Washington, DC: US Atomic Energy Commission. Available at: https://archive.org/details/0418_Plowshare_09_00_47_00.

Broad, W. 1992. *Teller's War: The Top-Secret Story Behind the Star Wars Deception.* New York: Simon and Schuster.

FitzGerald, F. 2000. *Way Out There in the Blue: Reagan, Star Wars, and the End of the Cold War.* New York: Simon and Schuster.

Hamilton, C. 2013. *Earthmasters: The Dawn of the Age of Climate Engineering.* New Haven, CT: Yale University Press.

Kaufman, S. 2012. *Project Plowshare: The Peaceful Use of Nuclear Explosives in Cold War America.* Ithaca, NY: Cornell University Press.

Kirsch, S. 2005. *Proving Grounds: Project Plowshare and the Unrealized Dream of Nuclear Earthmoving.* New Brunswick, NJ: Rutgers University Press.

Lindsay-Poland, J. 2003. *Emperors of the Jungle: The Hidden History of the U.S. in Panama.* Durham, NC: Duke University Press.

Masco, J. 2015. "The Age of Fallout." *History of the Present* 5 (2): 137–68.

O'Neill, D. 2007. *The Firecracker Boys: H-Bombs, Inupiat Eskimos and the Roots of the Environmental Movement*. New York: Basic Books.

Teller, E. 1960. "We're Going to Work Miracles." *Popular Mechanics* 113 (3): 97–103.

Teller, E., W. K. Talley, G. H. Higgins, and G. W. Johnson. 1968. *The Constructive Uses of Nuclear Explosives*. New York: McGraw-Hill.

Waters, C. N., et al. 2016. "The Anthropocene Is Functionally and Stratigraphically Distinct from the Holocene." *Science* 351 (6269). DOI:10.1126/science.

The Manual Pesticide Spray Pump

Michelle Mart and Cameron Muir

The chemical weed killers are a bright new toy. They work in a spectacular way; they give a giddy sense of power over nature to those who wield them, and as for the long-range and less obvious effects—these are easily brushed aside as the baseless imaginings of pessimists.
Rachel Carson, *Silent Spring* (1962)

If we accept Rachel Carson's metaphor that chemical weed killers are a bright new toy, the same is surely true of the ubiquitous device once used to dispense them: the manual pesticide pump. The manual pump felt like one of those water squirters that children frolicked with in sun-drenched backyards. It was easy to use, almost carefree—though it had a serious purpose: to kill insects, weeds, and other pests. The pesticide pump straddled the line between toy and weapon, especially as it gave the user the "giddy sense of power" described by Carson. The pump may have been a humble instrument, but it nevertheless illustrates the hubris of a culture convinced it can launch—and win—a war against nature, in this case by chemically manipulating it (plate 1).

The manual pesticide pump tells us much about the Anthropocene: the proliferation of novel, human-made compounds such sprayers were designed to disperse; the intimate ways in which materials flow between ecologies and our bodies; the unequal geographic and intergenerational distribution of risks; and the kinds of desires, visions, and political economies of the societies that made and used such technologies. The scale of the Anthropocene draws us toward the causes, legacies, and representative objects that are proportionately grand. The diminutive, bicycle pump–sized pesticide sprayer, on the other hand, directs our thinking away from the immensity of the Anthropocene and its abstract futures, down to the local, the present, and the everyday: to backyards, school playgrounds, neighborhood reserves, household kitchens. The participants in the Anthropocene are not just engineers or global mining companies, but ordinary folk, such as conservation volunteers, weekend gardeners, and parents concerned with household hygiene, health, and cleanliness. The pump itself played a historical role in making some of the disruptive processes that define the Anthropocene—in this case, industrial chemical production—become familiar and unthreatening. The physical pump helped to domesticate and naturalize the liquid poison that it contained, leading people throughout the world to accept—or even embrace—the idea that pesticides were integral to modern life.

* * *

The humble pump with the outsized impact grew in popularity following the launch of Flit® insecticide in 1923 and then its marketing in the easy-to-use "Flit gun." Other countries had their popular brands similar to Flit, such as Mortein, Baygon, Raid, and more that have since disappeared. The US fossil fuel company Standard Oil owned Flit and the pesticide's original formulation was based on mineral oils. Here we see the intricate interconnections and dependencies between fossil fuels, agriculture, and pesticides. Diesel-powered machinery enabled farmers to increase the scale and reach of their operations, but expansive monocultures would not have been biologically possible without petrochemical fertilizers and pesticides. Beginning in 1928 Flit, in its handy dispenser, was spread far and wide by the whimsical drawings of Theodore Geisel, an ad pitchman long before becoming the world-famous children's author Dr. Seuss. The tagline of Geisel's cartoon ads, "Quick, Henry, the Flit," continued to be used through the 1930s and 1940s, and moved beyond the pesticide ads to song lyrics and stand-up comedy routines. Soon Flit and the Flit gun became the generic names for all sorts of insecticides easily available to individual consumers to squirt wherever they wished. There is a great irony that Geisel, author of such mid- to late twentieth-century classics as *If I Ran the Zoo* and *The Cat in the Hat*, had earlier worked for Standard Oil and used such whimsy to sell Flit. Are pesticides whimsical? Maybe in the drawings of Geisel, but not when we consider how they affected the bodies of his young readers.

Had the appeal of the manual pesticide pump rested merely on a catchy phrase and childlike cartoons, its impact would have been limited, and it would not have earned its place in a pantheon of objects that exemplify the Anthropocene. Following World War II, though, the pump secured its significance in history when owners of Flit added the miraculous DDT to its formulation. By itself, DDT and its chlorinated hydrocarbon cousins that soon followed opened up a new age in which Americans and others far and wide aspired to the chemical control of the earth. But the marriage of DDT and the manual pump magnified the impact of both, and made them together greater than the sum of their parts. The pump individualized and commodified chemical power. Dominion over one's own private patch of earth could be had for the cost of a cheap tool filled with liquid poison from the local hardware store. Millions of people in the postwar world kept a supply close at hand. We recall our own families' under sink cupboard or garage shelf with a pump or spray bottle at the ready. Our families tell us stories about the pesticide fog truck rolling through suburban neighborhoods to eliminate nuisance mosquitoes. But more than taming the unruly nature that threatened the carefully planned order of suburban backyards, the pump imparted a sense of power and normality that paved the way for the acceptance of the commercial use of chemicals, sprayed over thousands of acres by airplanes and large tractors, not just by simple hand pumps.

It may seem like a long road from the manual spray pump of the early twentieth century to the widespread commercial use of synthetic pesticides that began in 1945. But the connection makes more sense when we focus on the demonstrable successes of pesticides and the idea that they could be used equally by individual homeowners and farmers growing food for a world market. The first indication that synthetic pesticides gave humans an unprecedented power over the natural world came in World War II when DDT conquered dangerous outbreaks of typhus, saving both soldiers and civilians. Soon after the war, when malaria remained a dangerous threat in many regions of the world, DDT appeared to offer a miraculous solution. Using the manual spray pump around one's own home allowed you to feel that same sense of power over insect pests and to embrace the use of pesticides in different situations.

In addition to worries about insect-borne diseases, many also worried about whether or not the world food system could feed a growing population, especially with ruined agricultural fields and factories destroyed by war. Newly industrialized agriculture—including the introduction of technologies such as synthetic fertilizers and pesticides—seemed to provide the answer for adequate food production. The amount of grain, dairy, and animal production on farms skyrocketed, dwarfing all previous output. Looking at such results, pesticides (starting with insecticides, but soon to include herbicides, fungicides, nematocides, and other poisons) appeared to be a magic bullet, and as safe as the formulations that individuals could spray from their own hand-held pumps. Ensuring adequate food production also had political implications in a Cold War world. American and other Western policymakers believed that countries that faced food and other shortages would be vulnerable to communist influence. Thus, pesticides, an example of superior modern technology, would help produce food and, at the same time, address a political threat.

Still, the question looms: why such profligate use of these chemicals—everywhere—apparently heedless that there might be any environmental consequence to such a choice? One answer is found in the stories that we tell ourselves—and that are told to us by companies and officials trying to direct our course. More than fifty years ago, Rachel Carson argued that many of us were blissfully ignorant of pesticides' power in the face of such messages: "Lulled by the soft sell and the hidden persuader, the average citizen is seldom aware of the deadly materials with which he is surrounding himself" (1962, 174).

Newspapers and magazines, in their articles and advertisements, offered enthusiastic reasons as to why pesticides should be used widely and liberally. Certain messages and images were repeated often throughout the 1940s and 1950s, providing a firm foundation for increasing pesticide use into the next century. Stories and ads contained numerous descriptions of how the chemicals increased output and killed their targets. Using scientific knowledge and new technology, pesticides demonstrated the essence of modernity, as Dow Chemical told readers of the *Saturday Evening Post* in 1951, "modern farms depend on modern knowledge."[1] Many ads asserted that pesticides were not only modern, but they were also safe, effective, and would make the world more beautiful. The herbicide "Weedone," for example, would bring "new, undreamed-of beauty! Spray away weeds easily, quickly, safely." Who could resist such promises? Such reassurances invited readers to use pesticides; they *domesticated* the chemicals. What better way to demonstrate domestication than to show pesticides at the easy disposal of housewives armed with manual pumps and, later, other simple packages, working to improve the lives of their husbands and children?

Although the articles and advertisements were exercises in domestication and normalization, they continued to echo the military rhetoric of World War II to frame the apparently never-ending struggle (war!) against pests. Unlike most wars, though, to be waged by experts and the military, this battle was to be fought by civilians as well. The pesticide spray pump was an appealing weapon against the relentless onslaught. Effectiveness against a persistent foe? Ease of use? Safety? All of these appeals promised heady success. They promised control over the natural world. Americans and other Westerners were eager to share the gospel. Thus, the messages about pesticides did not just domesticate chemical pesticides; they soon internationalized them as well through ambitious programs to export technology and thereby industrialize world agriculture. Such initiatives were not only about controlling nature, they were also about shaping the world through ideas of Western modernization and development. Union Carbide captured the spirit of this mission in a now-iconic 1962 advertisement showing a large, God-like hand pouring an elixir from a laboratory flask onto an agricultural field where a near-naked Asian Indian pushes a wooden plow behind a team of oxen; "India needs the technical knowledge of the Western world," the copy tells us.[2]

The same year that Union Carbide told readers that pesticides would bring miracles to the far reaches of India, Rachel Carson told her Western readers that such miracles have their costs, including insect resistance, harm to wildlife, rising expense, leaks of acutely toxic chemicals, and growing evidence of damage to human health. Years before Carson's book, starting in the late 1940s, some biologists warned about the environmental effects of the new synthetic pesticides, and the American Medical Association formed a committee to study their effects on human health. These early warnings, though, were eclipsed by the widespread adoption of the chemicals and the difficulty of finding direct evidence of harm outside of acute poisonings. Over the next few decades, it was difficult to isolate the single cause of a disease or condition in experiments that could be reproduced in the laboratory. Moreover, the quest to expose possible harmful effects of pesticides was less compelling than the chemicals themselves, which carried the illusion that with them we had the power to shape and improve nature.

To be sure, a lax regulatory atmosphere, a presumption in favor of industrial agriculture, and fierce defenders in the chemical industry have made it hard to provide definitive proof that a particular chemical caused a particular disease. Nevertheless, there have been hints and even clear signs in recent years that pesticides have had lasting health effects and thus symbolize the havoc wreaked by human ambition. A few examples will illustrate. The enzymes that pesticides target in pest species are identical or similar to enzymes in humans and other animals; thus, they disrupt how human enzymes function. Pesticides can cause genetic mutations, chromosomal alterations, and DNA damage. And, as new research in epigenetics is revealing, pesticides can also affect how cells express genes. Cancers and degenerative diseases associated with pesticides include leukemia, sarcoma, lymphoma, brain cancer in children, and Parkinson's disease. Pesticides can cause impairment of immune system function, social adaptability, reproductive health, and cognitive and physical development. In 2014 alarmed researchers for the medical journal *Lancet Neurology* described the situation as a "global, silent pandemic of neurodevelopmental toxicity." In a very concrete way, then, the pesticides that have saturated the environment for decades have also invaded our internal environment and have been absorbed by our bodies to change human health in the present—and in the future. Pesticides and other toxics that are part of us may seem at times invisible, but their impact is real and increasingly unavoidable.

The global crisis of chemical exposure is relatively recent. Since World War II, over 100,000 chemical compounds—including pesticides—have been released on the earth, only a fraction of which have ever been tested for safety. It is a profound change, one that can't be undone easily and not within our lifetimes. It will persist in the geological record and in our genetic legacies. This has occurred in a remarkably short time span. For this reason, pesticides are perhaps one of the least understood agents of the Anthropocene.

Take, for example, trying to figure out "safe" thresholds below which the impact of pesticides should be negligible. In 2009 a group of scientists from Europe and Australia introduced the concept of "planetary boundaries," a way to understand and measure the point at which human activity would breach earth systems resulting in sudden environmental change. One of the boundaries they discussed was chemical pollution, but they failed to quantify safe limits. One reason for their failure to do so was a lack of sufficient data for this under-studied phenomenon as well as the difficulty of tracing the spread of the chemicals. Pesticides are distributed by rivers, groundwater, soil, air, and food, and applied at different rates and simultaneously with other chemicals; in addition, their persistence rates are variable. This confusion about the spread of pesticides has implications for human health, since it is difficult to isolate and evaluate exposures; moreover, it turns out that individual genetics appear to make some people more susceptible to their effects.

It's clear that along with a culture that has long embraced pesticides, continued use of the chemicals rests on conflicting scientific advice about whether the chemicals are more harmful or helpful. In addition, the risk/benefit framework used to assess such advice is itself problematic. Who decides acceptable risk or desirable benefits? Is it manufacturers who profit from the pesticide use, farmers anxious to increase yields, doctors who spend their lives trying to cure cancer, or workers facing daily exposure? The debate may be contentious, but its results have been predictable. Risks and benefits are usually defined in economic terms, even if human and environmental health defies such categorization. The tragic consequences of choosing this economic risk/benefit paradigm are illustrated in the lives of poor farmworkers throughout the world.

In the late 1990s anthropologist Elizabeth Guillette asked children from the Yaqui Valley in Mexico to draw simple pictures—one group was from the agricultural lowlands where export crops for the global market were grown, the other from the pesticide-free foothills (Guillette et al. 1998). The children from the agricultural region could barely form shapes. The difference in the drawings is a striking visualization of what is largely an invisible agent of harm—and an illustration of the geographic inequality of toxic burdens (see figs. 1 and 2). Such inequality forces us to question how pesticide use became so widespread and so accepted. The manual pump—simple to use and resembling a toy—helped to domesticate pesticides decades before. By the end of the century, the implications of such domestication were clear: the chemicals were now shaping the lives of poor children and reaching down into the next generation, as cheap food was being traded for the lives of children.

Claims of ignorance continue even in the face of such evidence. Thus, we are left to ponder our embrace of pesticides. As we have argued, it is not based solely on material motivations in the drive to industrialize agriculture the world over. Interwoven with measurable factors of profit, loss, and pest numbers are naturalized assumptions about environmental control and human agency. Such assumptions, in turn, rest on a rich history of technological development and triumphant stories about the human relationship with the environment.

The simple pesticide pump encapsulated human ingenuity and technological prowess, and encouraged ever greater confidence in the ability to control the natural world with chemicals. Individual use of the manual pump helped to reinforce the idea that similar use of pesticides more broadly over fields and roadsides was desirable and necessary. The pump, thus, encouraged us to welcome industrial chemicals into our lives, normalized them, and gave power to a positive narrative of technical progress. The often unquestioned commitment to pesticides grew, heedless of the consequences for the children of the Yaqui Valley of Mexico, for the earth, and for us all.

Figures 1 and 2. Drawings produced by children from the agricultural lowlands and pesticide-free foothills of the Yaqui Valley in Mexico. Elizabeth Guilette, "An Anthropological Approach to the Evaluation of Preschool Children Exposed to Pesticides in Mexico," *Environmental Health Perspectives* 106 (1998): 347–53.

NOTES

1   Ad # 1, box 7, Dow Advertisements, Post Street Archives, Dow Collection, Othmer Library (Othmer), Chemical Heritage Foundation (CHF).

2   This ad first appeared in *Fortune* in April 1962 and was reprinted widely following the Bhopal disaster in December 1984.

BIBLIOGRAPHY

Carson, R. 1962. *Silent Spring*. Greenwich, CT: Fawcett Publications.

Guillette, E. A., M. M. Meza, M. G. Aquilar, A. D. Soto, and I. E. Garcia. 1998. "An Anthropological Approach to the Evaluation of Preschool Children Exposed to Pesticides in Mexico." *Environmental Health Perspectives* 106:347–53.

Kinkela, D. 2011. *DDT and the American Century: Global Health, Environmental Politics, and the Pesticide That Changed the World*. Chapel Hill: University of North Carolina Press.

Nash, L. 2006. *Inescapable Ecologies: A History of Environment, Disease, and Knowledge*. Berkeley: University of California Press.

Oreskes, N., and E. Conway. 2010. *Merchants of Doubt: How a Handful of Scientists Obscured the Truth on Issues from Tobacco Smoke to Global Warming*. New York: Bloomsbury Press.

Hubris or Humility?

Genealogies of the Anthropocene

Gregg Mitman

Hubris and humility. They are perhaps the two most common emotional responses to the Anthropocene. The first charts an environmental future of the "good Anthropocene," where technoscience provides the innovative tools for fixing a warming planet. The second propels us to a more dystopic environmental future, or at least a future filled with uncertainty, loss, and mourning in the face of accelerating species extinction and a world increasingly divided by those who have the means to survive and those who do not.

Hubris and humility, I suggest, are deeply rooted in the genealogies of scientific knowledge that have given birth to the Anthropocene, which may well be on its way to becoming a scientific object, given tangible material form in the strata of Earth's history. But why now? Why has the Anthropocene suddenly become the subject of scientific meetings, academic conferences, museum exhibits, journals, and popular articles? Surely, ever since Bill McKibben sounded a popular alarm in *The End of Nature* that global warming had marked a threshold in which nature was no longer, in his words, "an independent force," and that "by changing the weather, we make every spot of earth man-made and artificial," the idea of humans as a force of planetary change is hardly news (1989, 58, 65). Nor has it been for some time. But McKibben's emphasis on a threshold crossed, published the same year as the fall of the Berlin Wall, is suggestive of how the Anthropocene, and particularly the knowledge disciplines that sustain it, has its origins in the Cold War and its ensuing collapse. That same year, Francis Fukuyama would publish in *The National Interest* his similarly famous essay, "The End of History" (1989). "What we may be witnessing," Fukuyama wrote, "is not just the end of the Cold War, or the passing of a particular period of post-war history, but the end of history as such: that is, the end point of mankind's ideological evolution and the universalization of Western liberal democracy as the final form of human government" (4).

For a concept and potentially a scientific object that is meant to include humans as a geological force on a planetary scale, recognizable in the strata of deep time, the Anthropocene is remarkably resistant to considerations of its own historical genealogy. Like the 1989 writings of McKibben and Fukuyama, the Anthropocene trades in talk of rupture—it is an alleged rupture in scale, spatially and temporally, of the impact of the human species on earth. It reinforces a sense of novelty of the human species as a geological agent, either reveling in this new-found power of the human species in changing the face of the earth or crying out in an elegiac mode of loss and despair. In their essay, "Was the Anthropocene Anticipated?," Clive Hamilton and Jacques Grinevald dispense with any talk of historical precursors. "The Anthropocene," they insist, "represents a radical rupture with all evolutionary ideas in human and Earth history" (2015, 59). Hamilton and Grinevald reject any attempts to locate the Anthropocene concept in previous eras, whether it be the idea of the "biosphere" put forth by the Russian biogeochemist Vladimir Vernadsky in the 1920s or the notion of the "noosphere" advanced by the Jesuit priest and paleontologist Pierre Teilhard de Chardin in the early twentieth century. The difficulty, Hamilton and Grinevald argue, is that in such versions, "civilized Man emerges as a geological force *incrementally* over deep time." But such views of incremental human impact, they argue, obscure "the suddenness, severity, duration and irreversibility of the Anthropocene leading to a serious underestimation and mischaracterization of the kind of human response necessary to slow its onset and ameliorate its impacts" (67).

Ruptures allow little place for history; their power lies in unprecedented events. But the discourse of rupture is itself historically contingent. Ruptures surfaced in two critical historical moments in the genealogy of the Anthropocene: the dropping of the atomic bomb and the end of the Cold War. Both were transformative in reconstituting the disciplinary spaces of environmental knowledge, whereby the geosciences would competitively displace ecology in the scramble for who claimed "the environment" as its subject. Hamilton argues similarly that a "gulf . . . separates Earth system science from classical ecology, one that requires a leap from 'ecological thinking—the science of the relationship between organisms and their local environments—to Earth system thinking, the science of the whole Earth as a complex system beyond the sum of its parts.'" But I strongly disagree that the concept of the Anthropocene represents such a radical break from the past that history doesn't matter.

History matters a great deal. The very structure of knowledge that gave birth to the Anthropocene, built upon models of rupture and planetary crisis, has a specific geography and history in its production. For the Anthropocene is an object constituted through the Cold War nuclear arms race, which yielded unprecedented funding for the earth sciences and enabled, as Joseph Masco (2010) argues, "new public fears and visions of planetary threat" (2010, 9). Rupture and apocalypse, in addition to the hubris of geoengineering, were built into the scientific apparatus of the national security state. Remarkable continuities—in scientific personnel, computing, and future imaginaries—persisted as global warming replaced communism as the new planetary threat in the aftermath of the Cold War. Such Cold War histories are embedded in the *Plowshare* film, pesticide pump, and marine satellite tags, among other objects on display in the Cabinet of Curiosities for the Anthropocene contained in the pages herein.

But where sits ecology in the Anthropocene? A literature review of the top ten ecological journals based on impact factor suggests that the ecological sciences have been slow to take up the term and continue to shy away from critical engagement with the Anthropocene. This, too, is a contingency of history. In the rapid ascendance of planetary earth science, and the subsequent displacement of ecology as the sine qua non of the environmental sciences, we risk losing sight of life, in all its diverse forms, both human and nonhuman, that have shaped the planet. We concern ourselves with the Great Acceleration, failing to acknowledge that we are not the only species that have transformed the biogeochemistry of the earth. Cyanobacteria claim precedent by almost 2.5 billion years. Their sedimentary remains in stromatolites provide evidence of the Great Oxygenation Event, when their photosynthetic capacities transformed the atmosphere of the earth leading to widespread geological and biological change. It is hubris to suggest that we are the only species that has reshaped life on earth.

So let us step back in time for a moment, not into deep time, but into our more recent past, when ecologists willingly challenged the anthropocentric bias of their geological brethren. In the search to locate historical precursors of the Anthropocene, it is rather odd that the name Thomas Chamberlin has yet to appear on the list of scientific forerunners in the late nineteenth and early twentieth centuries. Chamberlin was a luminary in geology and unlike other candidates, such as Vernadsky and Teilhard de Chardin, his adoption of the Psychozoic, suggesting the age of man as a stratigraphic era, *did* make its way into discussions of past geological ages in a number of geology textbooks in the early twentieth century. Chamberlin was, as he liked to say, born on a moraine. The anecdote spoke to the long-lasting impact of his geological work on Pleistocene glaciations, which earned him the first directorship of the US Geological Survey's Pleistocene Division. Chief geologist and head of the Wisconsin Geological Survey from 1876 to 1882, at a time when hard rock mining in northern Wisconsin for iron ore was an economic backbone of the state and vital to a growing national steel industry, Chamberlin represented a long tradition in which stratigraphy, or classification of rock formations, was critical to the extraction of coal and minerals that spawned the Industrial Revolution. Indeed, the French comparative zoologist Georges Cuvier created the first geognostic map, detailing the layered structures of the earth around Paris using fossils as distinguishing markers. He did so in partnership with Alexander Brognigart, a mineralogist and director of the state porcelain factory outside of Paris. William Smith, an English mineral surveyor and canal builder, who coined the word "stratigraphy," used outcrops of distinct formations and observations of fossils commonly found within them to create a geognostic map of England and Wales, around the same time that Cuvier did, leading to bitter rivalry claims. Smith's maps are themselves wonders; Smith, however, died penniless, burdened financially by the extravagant costs of producing his enormous maps.

In the early years of stratigraphy in the mid-nineteenth century, national geological surveys paid particular attention to the Carboniferous group, aiming to provide as detailed and accurate information as possible about the outcrops of coal measure strata. The growth of stratigraphy was thoroughly entwined in the economy and geopolitics of coal. Yet the intertwined histories of stratigraphy and the dawn of fossil fuel extraction are amazingly absent from discussions of the Anthropocene. Such historical silences seemingly absolve the geological sciences of any responsibility for the unleashing of carbon into the atmosphere at unprecedented rates.

But Chamberlin's interests went well beyond those of economic geology and stratigraphy. Together with his former student, geographer Rollin D. Salisbury, Chamberlin advanced the study of physiography, which attempts to explain present-day landforms on the basis of past geologic process such as glaciation, erosion, and deposition. In his four-volume survey of the geology of Wisconsin, prepared with the assistance of Salisbury and published in 1883, Chamberlin introduced the Psychozoic era, representing the geology of the living present. "If the distinguishing of this as a new era is simply a recognition of the superior mental attributes of man," Chamberlin wrote, "the propriety of the classification may be fairly questioned; for, however pre-eminent man's intellectual and moral nature, as compared with the organisms that characterize earlier geological ages—however much man may transcend the Mammals, Reptiles, Fishes, and Invertebrates of the preceding eras, unless that superiority—or man, its working embodiment—is an efficient *geologic* agent, it does not entitle him to special recognition in a *geological* classification" (1883, 299).

Chamberlin believed man had acquired such a status on a strictly geological basis. Cultivation of the soil, along with the excavation and movement of materials, had, Chamberlin argued, altered physiographic processes of erosion and deposition. "The entire land life is being revolutionized through man's agency," he wrote. "That he will ultimately modify in a considerable degree marine life scarcely admits of question." Chamberlin argued that the domestication of plants and animals with the beginning of agriculture constituted the first epoch in the Psychozoic era. All previous periods of human existence, he maintained, belonged in the Cenozoic era, for they did not "greatly affect the course of geologic growth" (300). Two decades later, in a book that one geologist has described as "probably the most influential textbook of geology in the United States prior to World War II" (Dott 2006, 31), Chamberlin and Salisbury ended with a section on "prognostic geology," in which they predicted that a "Psychozoic era, as long as the Cenozoic or Paleozoic, or an eon as long as the cosmic and biotic ones" would likely "prove true" (Chamberlin and Salisbury 1906, 543).

I draw attention to Chamberlin not to construct an alternative origin for the history of the Anthropocene or to tell a Great Man story in the history of science. Far from it, although his location of a geological "age of man" in the Neolithic agricultural transition is of interest in light of contemporary debates about where exactly to date the Anthropocene, which hinges on a debate about the nature of change in reconciling earth and human history. It is also worth mentioning that Chamberlin was one of the first, along with Svante Arrhenius, to unite atmospheric chemistry with geology ("the ocean is an atmosphere in storage," wrote Chamberlin) in positing the carbon cycle as a principal driver of global climate oscillations in his search for causal explanations to account for periods of glaciation and retreat.

Far more significant, however, were different disciplinary reactions to the Psychozoic era. Such reactions are significant because they give us a clue into the ethical relevance and moral tales that the geological and biological sciences have inferred at different moments in time in their reading of earth's history, life's history, and human history.

In 1926 Edward Berry, a leading paleobotanist at Johns Hopkins University, wrote a rather scathing, tongue-in-cheek note in *Science* denouncing the Psychozoic era. The objects that garnered Berry's entry into the field of science were fossilized plants, but it was his knowledge of the living, not the dead, that made him look quite differently upon a geologic era that singled out humans as the dominant force on earth. As his National Academy of Sciences biographer noted, Berry's "keen appreciation of the meaning of fossil plants led him to see forests and prairies, coastal swamps and steaming jungles, where most geologists saw merely fossil leaves" (Cloos 1974, 64–65). He was capable of seeing, in other words, the ecological relations of living forms in the distant past.

Such engagement with other nonhuman species, grounded in ecology and botany, led Berry to question the rather anthropocentric-driven narrative of the geological history of the earth put forth by Chamberlin and Salisbury. "It is probably good philosophy to commence earth history with a hypothetical Archeozoic era, but is it equally good philosophy to terminate earth history with a Psychozoic era?" asked Berry. "No one would probably gainsay the magnitude and multiferous effects of human activity, but these are scarcely of geologic magnitude, and I can conceive of many past events as being of much greater importance than the advent of man, if viewed with a certain degree of detachment. Such, for example, as the origin of life itself. . . . It might be conceivable that the first mammal or the first flowering plant (Angiosperm) was more of an event than the first man" (1926, 16).

Berry became even more apoplectic. "It seems to me that a Psychozoic era is not only a false assumption, but altogether wrong in principle, and is really nurtured as a surviving atavistic idea from the holocentric philosophy of the Middle Ages," he bemoaned. At stake was the centrality of the human species in the long evolutionary history of life on earth. While "there can be no objection to speaking of the present as the Age of Man—or Woman—for that matter," Berry concluded, that was a far cry from establishing a formal geologic era in honor of the human species. Berry insisted that there was no stratigraphic evidence or reason for doing so.

Berry's essays appeared during a period in the history of the life sciences when ecology was in ascendance, and when attention to relationships between organisms and their environments brought forth novel experiments in trying to understand and represent the different life worlds of other beings with whom humans inhabit this earth. It is no coincidence that in the recent turn to multispecies ethnography and to a version of posthumanism informed by animal studies, scholars are resurrecting the work of biologists in the interwar years, such as Jakob von Uexküll or Karl von Frisch, in their efforts to make visible through a diverse array of ethnographic encounters across species divides the existence of parallel universes all around us, which are inhabited by beings living in different perceptual worlds in different scales of time. Whatever its multiple causes—the deadliest war and global pandemic in modern history, which, combined, killed an estimated 60 million people in a few short years; a global economic depression; or the flourishing of ecological and evolutionary science—claims to the superiority of the human species over other forms of life rested on shaky grounds.

And it generated new forms of historical writing in which other species displaced their human companions from center stage. Consider, for example, Hans Zinsser's *Rats, Lice, and History*, originally published in 1935. A bacteriologist and gifted writer who served on the American Sanitary Commission during the Great War in an effort to combat a typhus epidemic raging on the front lines, which at its height resulted in 9,000 new cases arising each day, Zinsser set out to write a world history of the human species, from the perspective not of man, but of a microbe and an insect. Written as a biography of a disease, *Rats, Lice, and History* introduced an organism—typhus—and its host, the louse—as actants in history long before the emergence of actor-network theory or debates about the agency of nature within environmental history. Harnessing the tools of ecology and evolution, Zinsser set out to tell the "louse's point of view in its relationship to man" in writing "the biography of a protoplasmic continuity like typhus" (2008, 166). How vulnerable the human species became, how humble its triumphs looked, when considered from the viewpoints of companion species across the spans of ecological and evolutionary time.

Fast-forward three human generations to another pairing of the biological and earth sciences. The year is 2000. Our first author is a Dutch engineer, who got his start in science as a computer specialist at the University of Stockholm's Meteorological Institute, modeling the effects of nitrous oxide on stratospheric ozone. Our second is a paleoecologist who spent his professional career exploring the biology, ecology, and taxonomy of diatoms, those minute aquatic organisms responsible for every fifth breath you take and fixing more carbon than all the world's tropical rainforests. Together Paul Crutzen and Eugene Stoermer, in the pages of the newsletter of the International Geosphere-Biosphere Program, noted the "major and still growing impacts of human activities on earth and atmosphere" at all scales, including the global (Crutzen and Stoermer 2000, 17). Citing evidence that ranged from a tenfold increase in human population over the last three centuries, to the rapid exhaustion of fossil fuels "generated over several hundred million years," from the substantial increase of nitrous oxide, carbon dioxide, and methane into the atmosphere to the significant depletion of primary production in the world's oceans through human predation, Crutzen and

Stoermer proposed using the term "anthropocene" to denote the current geological epoch in which we now live and to emphasize the "central role of mankind in the geology and ecology" of the planet. Crutzen and Stoermer came from two quite different lineages of postwar environmental science, as the objects with which they identified and brought them into their respective fields of geophysics and ecology suggest. The stuff of their science—atmospheric gases and models, diatoms and lake sediments—are also enmeshed in different but overlapping infrastructures that have shaped the contours of the Anthropocene, its affective registers, as well as its imagined pasts and futures.

A glimpse into those differences can be found just two years later, when Crutzen in the pages of *Nature* introduced the concept of the Anthropocene once again, this time without Stoermer. While the essay covered much the same ground as the coauthored IGBP piece, there are notable differences. A number of the biological examples Crutzen and Stoermer had cited as evidence of escalating human impact, including destruction of coastal wetlands and the altering of geochemical cycles of freshwater biotic communities, are notably absent. More significant, however, is the concluding paragraph in *Nature*. To "guide society towards environmentally sustainable management in the era of the Anthropocene," Crutzen concluded, "may well involve internationally accepted, large-scale geoengineering projects, for instance, to optimize climate" (Crutzen 2002, 23).

Crutzen has been much more hesitant in recent interviews about "techno-fixes" to global warming. Indeed, science journalist Christian Schwägerl argues that Crutzen's experiences with chlorofluorocarbons—Crutzen was one of the key scientists to make visible their destructive impact on the ozone layer—"has made him humble in the face of earth's complexity" (Schwägerl 2014, 221). Whatever Crutzen's individual ethical stance toward engineering the planet may be, the note of technocratic optimism on which he ended his *Nature* article is symptomatic of one strand of futures thinking inherent in the rise to prominence of the geophysical sciences during the Cold War. Harnessing the forces of nature on a global scale in the interest of defense required knowledge about the earth's physical environment: its atmosphere, oceans, and lithosphere. Keeping a thumb on the pulse of the planet arose as a result of both American and Soviet interest in keeping a watchful eye on the bombs going off in each other's territories and radioactive fallout that blanketed the globe. The Cold War produced the scientific infrastructure, data, and research that would ultimately provide, as Paul Edwards (2013) and others have argued, the evidence for climate change as well as arguments for the birth of the Anthropocene.

But whose voices have been silenced, whose futures have been displaced, in narratives reliant upon global models that risk ignoring local ecologies: the resilience and vulnerability of different human populations who don't readily aggregate into a universal "we," and the primacy of other nonhuman species critical to the survival of our own? Indeed, it is disconcerting how quickly Stoermer's contribution to the concept of the Anthropocene has been forgotten with his death in 2012. It was the lifeworld of another species—diatoms—that opened Stoermer's eyes, like my own, to ways of being in the world quite distinct from the human-centered one in which I and my human kin live. Through the shifting distribution and abundance of algal species in the Great Lakes region, revealed through lake sediments, as well as an intimate understanding of their livelihoods and needs, Stoermer was able to tell a story of land-use change in the Great Lakes region in which humans, plants, and animals all had a part.

The marginalization of ecology in the rise of planetary-scale geoenvironmental sciences risks turning the Anthropocene into a Promethean narrative of human mastery and control. While work in the ecology and systematics of the microbial world increasingly reveals how we, as humans, are but an entangled bank, a complex assemblage of animal-micro-biome interactions, the Anthropocene strikes back with a vengeance, reasserting the primacy of *Homo sapiens* in driving the evolution of life, for good or ill, on the planet. Yet, such a viewpoint ignores how even the human genome is indicative of the interdependence and relationality of living forms that came together as partners in the changing development and evolution of humanity. We need a chorus of voices, from different knowledge disciplines, from people who occupy different places and walks of life on the planet, and from other nonhuman species to temper the hubris of the anthropos, as Berry recognized long ago in his critique of the Psychozoic era. Objects like snarge, hybrid corals, and the sounds of the Huia bird in this Cabinet of Curiosities speak powerfully to the way human histories are dependent upon the histories of nonhuman beings in their tellings.

While stratigraphers will debate and ultimately decide whether the Anthropocene officially marks a new geological epoch in the annals of science, we dare not cede discussion of its meaning and implications to those occupied solely with rocks, sediments, and chemistry. It is, after all, life on the planet—past, present, and future—that bears witness, in fossilized bones and living flesh, to large-scale anthropogenic change. As we think beyond the instantaneous to longer frames of time, and scale up environmental problems to the global level, we should be cautious to not lose site of the diversity of lives—human and nonhuman—differentially impacted by planetary change.

BIBLIOGRAPHY

Berry, E. W. 1926. "The Term Psychozoic." *Science* 64:16.

Chamberlin, T. C. 1883. *Geology of Wisconsin. Survey of 1873–1879*, vol. 1. Commissioners of Public Printing. http://digital.library.wisc.edu/1711.dl/EcoNatRes.v01.

Chamberlin, T. C., and R. Salisbury. 1906. *Geology*, vol. 3, *Earth History*. New York: Henry Holt and Co.

Cloos, E. 1974. *Edward Wilber Berry, 1875–1945. National Academy of Sciences Biographical Memoir*. Washington, DC: National Academy of Sciences.

Crutzen, P. 2002. "The Geology of Mankind." *Nature* 415:23.

Crutzen, P., and E. Stoermer. 2000. "The Anthropocene." *Global Change Newsletter* 41:17–18.

Doel, R. E. 2003. "Constituting the Postwar Earth Sciences: The Military's Influence on the Environmental Sciences in the USA after 1945." *Social Studies of Science* 33:635–66.

Dott, R. H., Jr. 2006. "Rock Star: Thomas Chrowder Chamberlin (1842–1928)." *GSA Today* (October): 30–31.

Edwards, P. 2013. *A Vast Machine: Computer Models, Climate Data, and the Politics of Global Warming*. Cambridge, MA: MIT Press.

Fukuyama, F. 1989. "The End of History." *National Interest* (Summer): 1–18.

Hamblin, J. 2013. *Arming Mother Nature: The Birth of Catastrophic Environmentalism*. Oxford: Oxford University Press.

Hamilton, C., and J. Grinevald. 2015. "Was the Anthropocene Anticipated?" *Anthropocene Review* 2 (1): 59–72.

Haraway, D. 2016. *Staying with the Trouble: Making Kin in the Chthulucene*. Durham, NC: Duke University Press.

Masco, J. 2010. "Bad Weather: On Planetary Crisis." *Social Studies of Science* 40:7–40.

McKibben, B. 1989. *The End of Nature*. New York: Doubleday.

Rudwick, M. J. S. 2014. *Earth's Deep History: How It Was Discovered and Why It Matters*. Chicago: University of Chicago Press.

Schwägerl, C. 2014. *The Anthropocene: The Human Era and How It Shapes Our Planet*. Santa Fe, NM: Synergetic Press.

Tsing, A. 2015. *The Mushroom at the End of the World: On the Possibility of Life in Capitalist Ruins*. Princeton, NJ: Princeton University Press.

Van Dooren, T. 2014. *Flight Ways: Loss and Life at the Edge of Extinction*. New York: Columbia University Press.

Zinsser, H. 2008. *Rats, Lice, and History*. 1935. Reprint, New Brunswick, NJ: Transaction Publishers, 2008.

Living

and Dying

Huia Echoes

Julianne Lutz Warren

There is nothing for you to say. You must / Learn first to listen . . . / And,
though you may not yet understand, to remember.
  W. S. Merwin, "Learning a Dead Language" (2005)

the huia-trapper // whistles the song / I try to resist // I want to tug /
something out of him // the radio voice says / *believed to be extinct*
  Hinemoana Baker, "Huia, 1950s" (2004)

The Object

This chapter's object—which embodies the Anthropocene—is an aural relic. This
relic is the recording of a human imitation of extinct birdsong, which I am calling "Huia Echoes." "Huia Echoes" is a dramatic chorus for our age, and beyond
(plate 4).

Prelude: First Encounter

A few years ago, I was searching the audio archives of the Macaulay Library of
the Cornell University Lab of Ornithology for recordings of living birds to accompany a talk on "Remembering Nature as Hope." In the process, I incidentally

came across the call of an ivory-billed woodpecker. I knew that this bird kind of the southeastern United States and Cuba was likely recently extinct. I caught my breath when I heard this vanished voice. My awareness roused, I made a list of the avian species listed as extinct by the "IUCN Red List of Threatened Species" and checked to see how many of these birds' songs and calls had been saved in Macaulay's collection. I discovered that of 140 extinct species, the voices of only 5 were represented. Hearing each one evoked poignant feelings. Catalogue number 16209 titled "Human Imitation of Huia"—a mid-twentieth-century soundtrack of a now-deceased Māori man mimicking songs of already extinct huia, a bird endemic to Aotearoa New Zealand—in particular, haunted me.

I could not forget these dead voices, living on.

May we never forget.

Perhaps more of us, following poet Merwin's advice to "Learn first to listen" —to this bonded group of singing remains—will also remember and come to deeper hearing. Perhaps, in hearing, as Baker in her poem writes, though we may "try to resist // . . . to tug / something out" of the multiplex voice, we will learn that something from within ourselves is wanted to help enrich and multiply the whistling echoes.

The Historic Score: "Human Imitation of Huia"

The recording in Macaulay Library titled "Human Imitation of Huia" includes narration by Robert Anthony Leighton Bately, a man of British stock, descended from pioneer families. He explains that what we are hearing is a Māori man named Hēnare Hāmana—a bird mimic who in his younger days had heard living huia—whistling his re-creation, after they were extinct, of a sonic scene. In this imagined plot, a male and a female bird carry on a dialogue as they feed together in a forest. Here is that historic recording with Bately's narration:

Audio 1: Listen to "Human Imitation of Huia": http://macaulaylibrary.org/audio/16209.

Figure 3 is the recording transcribed as a score.

Presenting the Object: "Huia Echoes," A Dramatic Chorus

The narration helps sketch the story behind the imitated birdsong in the original recording. It was the whistle that charmed me, though. So, with the generous help of technicians, we removed the narration, freeing only the song to replay (see fig. 4).[1]

Bateley: "Let us imagine two birds are feeding on a rotten tree. After awhile, the female climbs to the top of the tree and glides into the distance. The male bird calls with the following notes," whistled by Hamana:

Male call: 19 seconds

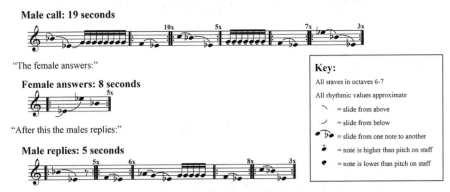

"The female answers:"

Female answers: 8 seconds

Key:

All staves in octaves 6-7

All rhythmic values approximate

⌐ = slide from above

⌐ = slide from below

●♭● = slide from one note to another

● = note is higher than pitch on staff

● = note is lower than pitch on staff

"After this the males replies:"

Male replies: 5 seconds

"The male bird then joins the female where they scratch among the leaves. After feeding they both climb to the top of a tōtara tree. And, now, to complete the survey of the huia, Mr. Hamana will repeat his calls. The male:"

Male repeat: 15 seconds

"The female:"

Female repeat: 8 seconds

Figure 3. Transcribed musical notation of whistled version of Huia songs with accompanying narration found in *Human Imitation of Huia*, catalogue number 16209 recording, Macaulay Library, Cornell Lab of Ornithology. Courtesy of Martin Hatch.

The intention of the descriptive words along with the human memory of a native bird tongue still shape the grammar of the musical phrases as the dramatic chorus resounds.

Audio 2: Listen to "Huia Echoes," a song of the Anthropocene: http://www .nzbirdsonline.org.nz/sites/all/files/27%20-%20Huia%20%28Imitation%29.mp3.

This, then, is the aural relic I am calling "Huia Echoes"—the chorus of extinct birdsong, echoed by human voice, echoed by machine, which may be played repeatedly—beginning, middle, end, beginning—looping into listeners' heads, potentially echoing on.

"Huia Echoes": Biographical Notes

Brief History of Huia, the Echoes' Source—Of all the lands of this vast earth, huia, a unique wattlebird, inhabited mainly the northernmost of a pair of stormy southern islands that rifted from Gondwana 80 million years ago. Huia ancestors may

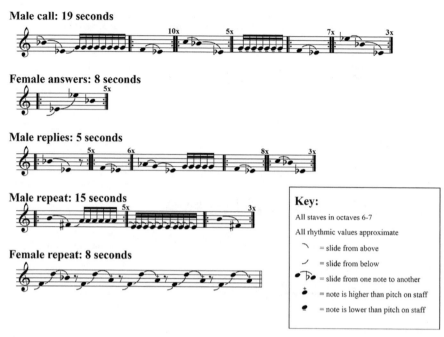

Figure 4. Transcribed musical notation of whistled version of Huia song extracted from *Human Imitation of Huia*, catalogue number 16209 recording, Macaulay Library, Cornell Lab of Ornithology. Courtesy of Martin Hatch.

have flown here from Australia on westerlies across the sea 50 million years later. The islands' first human beings, the Māori ancestors, finally appeared just 800 years ago. The name they gave the birds sounds like their song—*huia*. And the birds' place, also the people's new home, they called *Aotearoa*, or, in English, "long white cloud." Later, European colonists, whom Māori named Pākehā, christened the islands *New Zealand*. The bird, in Latin, became known as *Heteralocha acutirostris*, which in English means something like "the husband's is different from his wife's piercing sharp beak."

Huias' best-known calls have been described as a flute-like whistle with a prolonged note followed by short, quickly repeated ones, and as a recurring legato phrase quivering at the end. The birds' songs issued from their ivory bills, which were sexually dimorphic to an unusual degree. Females' bills were lancing-long and gracefully curving. Those of males were short and sharp like pick-axes. A pair of orange wattles, fleshy pendants ornamenting the gape flanges of both sexes, contrasted brightly with feathers that were silky blue-black from head to tail. The tips of a huia's twelve tail feathers, however, like his or her bill, were the color of ivory.

In the early decades of the twentieth century, huia joined a long line of these islands' birds—a quarter of them, or over fifty species—who have become extinct since the first human contact in the thirteenth century. More than half of these species, including every kind of moa, vanished between the time of Māori and eighteenth-century European arrivals. The rest were rapidly lost after Pākehā came. And, currently, many more species—including huias' closest relatives, saddleback (tieke or *Philesturnus carunculatus*) and North and South Island kōkako (*Callaeas wilsoni* and *C. cinerea*)—are on life's brink.

The loss of huia, extinct by the early twentieth century, can be blamed on a constellation of place-specific, human-initiated causes that today also ring, repeatedly, with global familiarity. Causes involved acute and chronic disruptions of long-evolved interdependencies among minerals, soils, waters, plants, animals, and air. At the time of Māori ancestral canoe arrivals, the islands' only mammals were bats. These first people brought with them bird-hungry Pacific rats. Then, a few hundred years later, European ships delivered more mammalian predators, like Norway rats, cats, stoats, and ferrets. Red deer from Scotland ate regenerating forest; and exotic birds, such as minas from India, brought unfamiliar ticks that stressed local birds.

Humans also dispatched huia directly. Traditionally, Māori hunters snared them for their beautiful tail feathers used for chiefly and sacred purposes. With the firepower of guns and the commodification of their feathers as hat ornaments (particularly after the future King George V donned one), and as parlor curiosities and museum specimens, Pākehā and Māori hunting intensified. From the nineteenth century, intense Pākehā-driven alterations of land and water also expanded. The new-come imperialists bought or appropriated wide swaths of forests and swamps, many of which were huia and Māori whenua or ancestral places, supporting and supported by interwoven avian-human indigenous identities. The newcomers burned, timbered, and drained these places and divided long-standing relationships in exchange for a managed system familiar to them— one of grass pastures, sheep and cows, and crops of potatoes, oats, and wheat, mined minerals and fossil hydrocarbons, railways and towns of well-warmed houses with weeded gardens, Chinese cherry trees, roads, shops and banks, stone cathedrals, museums, radio stations and recording machines.

Echo 1, Human Voice: Curious huia could be lured near to a practiced imitator whistling a resemblance to their songs. As a young man, Hāmana (b. 1880; d. >1949)[2]—a member of Te Aitanga-a-Māhaki, Ngāti Porou, known in Bately's words as "a local Māori experienced in giving huia calls"—assisted in at least two Pākehā-led huia search expeditions in 1908 and/or 1909. Only one bird was encountered on the earlier expedition through formerly prime habitat in the northern portion of the Ruahine Range of the North Island.

Huia had occupied wet mountain forests with arching tree branches of wide-girthed tōtara with gold-flaking bark, rendered by Pākehā artists as cathedral-like, and stands of southern beeches floored with decaying boles stocked with huhu grubs and hinau trees with tasty purple berries, both of which huia and Māori liked to eat. Huia frequented tangled manuka groves teeming with tree-crickets, another bird delicacy, on grounds sloping into brook-fed ravines of towering crimson-flowered rewarewa and pukapuka shrubs fragrant with cream-colored blooms. In the soundtrack, now as an aging man, Hāmana echoes a pair of remembered huia voices, whose kind no one will ever hear again in the flesh, singing to each other in an area of their former forest.

Echo 2, Machine Recording: The Pākehā habit of collecting skins of birds known to be endangered to save some museum knowledge of them, or to keep as cabinet curiosities, perhaps extending even to takings for keepsakes of Māori tradition, paradoxically, reduced avian numbers already in perilous decline. Recording equipment, on the other hand, could multiply rather than deplete stocks of avian songs, but was not readily available before huia were gone.

By 1949 the city of Wellington had a radio station with recording facilities. Understanding the bird to be an "object of unusual interest," Bately, as a local historian and author, wanted "to preserve a resemblance to the call of the huia . . . which is believed extinct." So Bately invited Hāmana, who, like him, lived in Moawhango near Taihape, to travel together about 140 miles south to station 2YA's (now RNZ National) studio.

There, prompted by Bately, Hāmana whistled his recollection of huia calls into a microphone. Technical experts used a recording lathe to etch the composite music of native bird tones and Māori echo, plus Pākehā narration, into a spiral of grooves on a black lacquer disc, which, as it spun in contact with a needle, could be played back. This machine sounding, then, is a second echo that not only reproduced a remnant of the extinct birdsong, but also saved human memories of huias' phrases, along with the thus-obscured cultural tradition of learning them. All of these losses were given a voice.

Echo 3 and Echoing On, Song-repeating Listeners: The bird-man-machine soundings thereafter circulated and multiplied into countless other echoes, in reproduction of the recording sung out by turntables and by newer kinds of playback machines, and, by some listeners, even embodied and rehummed into the living world. Soon after the Wellington recording was made, the dramatic soundtrack was presented as part of a talk on "Native Birds of Our District" by V. Smith of Taihape to the Royal Forest and Bird Protection Society, which appears to have held the phonograph record. Later, the original ten-inch acetate disk was copied onto tapes, including by the New Zealand Broadcasting Corporation. John Kendrick, a New Zealand conservationist, sound recordist, and radio host of "Morning Report bird calls," took a copy of their tape. This copy was copied by field collaborator William V. Ward for the Cornell Lab of Ornithology. The lab labeled the recording as catalog number 16209 in their Library of Natural Sounds, now the Macaulay Library. Macaulay began digitizing in 2000, subsequently making their holdings available to echo on with a quick click through the Internet. This, as I've explained, is how I first encountered "Human Imitation of Huia," which became edited into this chapter's focal object—a sonic artifact, which I am calling "Huia Echoes."

Spinning "Huia Echoes"

There is "a way the older people have of telling a story," Māori author Patricia Grace says, "a way where the beginning is not the beginning, the end is not the end. It starts from the center and moves away from there in such widening circles that you don't know how you will finally arrive at a point of understanding, which becomes itself another core, a new centre" (Thompson 2008, 66).

Perhaps "Huia Echoes" is telling this sort of story, starting at the core of a once-feathered source of destroyed-forest birdsong, circling out in a formerly-forest-bird-interwoven-man's voice, recorded by a descendant of colonist pioneers into the grooves of a spinning disc, then copied into other machines to repeat into air, potentially resounding through unknown ears and recurring in others' tongues elsewhere.

This choral artifact as a whole, then, might enchant our imaginations into another central starting place that begins with listening to "Huia Echoes" as a different kind of being. Indeed, this compound voice, I have come to feel, unexpectedly, is not an object after all. The extinct music somehow is not dead. Latent within technology, "Huia Echoes" is an alive companion, evident when I switch on a machine. Indeed, keeping near, housed in my iPhone, this musical storyteller by encouraging me to hear others helps me feel less alone.

Flowing through a legacy of saved memories—elemental, biotic, and mechanical—through a small speaker, the birdsong traces replay into different places. I begin to understand the mimicked dead birdsong as a de-feathered, skin-less teacher, an audible silence—a reverberating absence—bringing forward the past in moving conversation with the present.

For example, listening in boreal Alaska's Atigun Pass, I hear the colonist's machine-bound avian and human prisoners absorbed into wind sounding on rocks, water, and tundra leaves. I want to shout "Quiet!" to the play of air. But, keeping myself still, I also wish the currents to rush on in their forgetting way, dissipating cruelties to each unique winged-body and dark-skinned person who has suffered them. An inkling blows in from behind, whispering: we belong to each other.

As I listen in the foggy pillared peaks of Wulingyuan Scenic Area of China's Hunan Province, "Huia Echoes" pushes through a din of human-crowd voices so effectively that the whistle draws curious and also nervous looks. As do I, with my blue eyes and pale skin. My first impulses want me and my singing friend to hush or blend in alongside a contrary one to defend us both in a very loud voice, followed by an urge to announce my history of oppressing failures—personal and ancestral—to act with such spirited care toward all manner of life, accompanied by a humiliating feeling that this in itself can be self-aggrandizing. An insight rises from within, humming: desire healing.

It is this legacy of failure—institutionalized—that has delivered the world-of-life into a global epoch of dire consequences, still unfolding—many of which, despite anyone's deepest desire otherwise—can never be unmade, like huia's extinction—an entire bird language—extinguishing entwined Maori sacred tradition. This is the epoch that some have dubbed the Anthropocene, which might be considered yet another starting point for a fresh round of storytelling.

Anthropocene Remains

The Anthropocene, in albeit contested geological terms, is characterized by marks of worldwide human domination in fossil and chemical changes in soils, sediment, ice, or rock. In cultural terms this is an epoch of evidence-based perceptions of rippling, unintended outcomes of human actions reversing billions of years old trends of generative Earth. Reverses include unprecedentedly rapid rates of extinction—careening, in a matter of centuries, toward the likelihood of over 75 percent of bird species missing plus a similar proportion of other living types—with soil fertility diminishing faster than building up interpenetrating with global climate change, rippling in other forms ruin, unjustly distributed.

Injustices might not readily register in geological records but, humanely, it is clear that not all human beings are dominators. Many are unwillingly if not inculpably embedded in an imperial system—of intensive mining of lands and waters, toxic industries, including agriculture, and fossil-fuel burning—imposed by actors in a centuries-old Occidental narrative of "the enlarging of the bounds of Human Empire, to the effecting of all things possible," in Englishman Francis Bacon's words, joined with insatiable desire for wealth. This colonizing saga has notably benefited white-skinned people at the expense of those on the most vulnerable front lines of thus suffering Earth, including feathered, finned, and leafy, rooted beings, human ones who live within thin walls, and/or who are women, people in brown and black skins, children, inhabitants of coasts, small oceanic islands and African countries.

But the Anthropocene tale is not finished. The captive irony of this epoch, which "Huia Echoes" helps announce with widening understanding, is that the victory of human empire—a pyrrhic one—has become the unintended trappings of that same dominating intent. In other words, as the price of privilege has enlarged globally—extinguishing huia forest-singing, swallowing the voice of a Maori hunter into a machine alongside swelling hosts of others—mounting debts surround and undermine even the thickest walls of the imperial-minded.

The unfettering irony of the Anthropocene, on the other hand, is that world-wide political unrest and intensifying storms and droughts resulting from the overconfident efforts of dominating humans bare how all life interpenetrates with a still-wild planet fecund with music and inventiveness. The future may be darker than it ever was, our ignorance great, yet "Huia Echoes"—singing remains carefully saved in a machine—may be released in each of us who listen, learn, and remember this tugging voice, holding together a willing companionship of diverse others who together resist tugging. Here might be a new starting place of telling tales, not that everything will be alright, but of widening collaboration of human beings within an ecosphere of mutual belonging, the hope of healing.

NOTES

1   With thanks to Macaulay Library at the Cornell Lab of Ornithology, USA for permission to use "Human Imitation of Huia," Catalog #16209, William V. Ward, recordist, and to Collections Management Leader, Matthew Young, for technical help with editing.

2   With thanks to Kate Evans of *New Zealand Geographic* and Sarah Johnston of Ngā Taonga Sound & Vision for their correspondences, and for several helpful discoveries of new materials and historical details related to both Hēnare Hāmana and Robert Bately and the 1949 recording, which they also dated.

BIBLIOGRAPHY

Baker, H. 2004. *Matuhi/Needle*. Wellington: Perceval Press/Victoria University Press.

Barker, F. K., et al. 2004. "Phylogeny and Diversification of the Largest Avian Radiation." *PNAS* 101 (30): 11040–45.

Best, E. 1982. *Maori Religion and Mythology: Part 2: Origin of Birds*. Wellington: P. D. Hasselberg.

Buller, W. 1888. *A History of the Birds of New Zealand*, vol. 1. London: Published (for the subscribers) by The Author, 8 Victoria Chambers, Victoria Street, Westminster, S.W.

Cornell University. N.d. "Human Imitation of Huia. Catalog 16209." Audio Archive Recordings, Macaulay Library, Cornell Lab of Ornithology, Ithaca, NY. http://macaulaylibrary .org/search?media_collection=1&taxon_id=&taxon_rank_id=&q=16209.

Evans, K. 2016. "Echoes of the Past." *New Zealand Geographic* 139 (May–June). https://www .nzgeo.com/stories/huia.

Forster, M. 2014. "Indigeneity and Trends in Recognizing Maori Environmental Interests in Aotearoa New Zealand." *Nationalism and Ethnic Politics* 20:63–68.

Gurche, J. 2013. *Shaping Humanity: How Science, Art, and Imagination Help Us Understand Our Origins*. New Haven, CT: Yale University Press.

International Union for Conservation of Nature and Natural Resources (IUCN). 2016. "Red List of Threatened Species." http://www.iucnredlist.org.

Johnston, S. 2016. "Te Karangaa Te Huia/The Call of the Huia." Nga Taonga Sound & Vision. http://www.ngataonga.org.nz/blog/nz-history/the-call-of-the-huia/.

Kolbert, E. 2014. *The Sixth Extinction: An Unnatural History*. New York: Henry Holt and Company.

Licht, A. 2009. "Sound Art: Origins, Development and Ambiguities." *Organised Sound* 14 (1): 3–10.

McKibben, B. 2010. *Earth: Making Life on a Tough New Planet*. New York: Times Books.

Merwin, W. S. 2005. *Migration: New and Selected Poems*. Port Townsend, WA: Copper Canyon Press.

Orbell, M. 2003. *Birds of Aotearoa: A Natural and Cultural History*. Auckland: Reed Publishing.

Pawson, E., and T. Brooking, eds. 2013. *Making a New Land: Environmental Histories of New Zealand*. Dunedin: Otago University Press.

Phillips, W. J. 1963. *The Book of the Huia*. Christchurch: Whitcombe and Tombs.

———. N.d. "Huia Research: 1927–1954 (MU000235)." Dominion Museum (creating agency), Museum of New Zealand/Te Papa Tongarewa, Wellington.

Riley, M. 2001. *Maori Bird Lore*. Paraparaumu, NZ: Viking Sevenseas, NZ.

Swimme, B., and M. E. Tucker. 2011. *Journey of the Universe*. New Haven, CT: Yale University Press.

Tennyson, A., and P. Martinson. 2007. *Extinct Birds of New Zealand*. Wellington: Te Papa Press.

Thompson, C. 2008. *Come on Shore and We Will Kill and Eat You All*. New York: Bloomsbury.

Warren, J. 2011a. "Remembering Nature as Hope." *Newfound* 2 (1). https://newfound.org/ archives/volume-2/issue-1/non-fiction-warren/.

———. 2011b. "Urgent: Dreams." *Journal of Environmental Studies and Sciences* 1 (3): 256–61.

———. 2015. "Hopes Echo." Poetry Lab of the Merwin Conservancy, November 2. http://www. merwinconservancy.ort/2015/11/the-poetry-lab-hopes-echo-by-author-julianne-warren-center-for-humans-and-nature/.

Whatahoro, H. T. 2001. *The Lore of the Whare-Whanga: Or, Teachings of the Maori College on Religion, Cosmogony, and History*, vol. 1, *Te Kauwae-Runga, or "Things Celestial."* Translated by S. P. Smith. Cambridge: Cambridge University Press.

Snarge

Gary Kroll

This feather belonged to a Canada goose that summered in Labrador and migrated along the Atlantic flyway to winter somewhere in New Jersey most likely. On a cold winter morning, January 29, 2009, the goose was flying with a skein of conspecifics at about 3,000 feet. Just as it was passing over the Bronx, the goose—and many of its colleagues—collided with an airplane that had taken off from LaGuardia Airport just minutes before. This particular goose was "ingested" by the left engine, leading to catastrophic failure. A similar drama unfolded in the right engine and so the plane was forced to land in the Hudson River. All the humans were fortunate to survive. It was a "Miracle on the Hudson."

Richard Dolbeer of the USDA's Wildlife Services was on the team that investigated the plane once it had been drawn out of the river. There is no bigger name in the world of bird-strike mitigation; the drama of the event must have required the investigative power of a seasoned expert. He scraped what was left of the goose from the left engine's outer flowpath, put it in a plastic bag—designed for just such a purpose—made some notes on the bag, and sent it to the Feather Identification Laboratory at the Smithsonian. The naturalists at the lab quickly identified the species, *Branta canadensis*, notified the National Transportation Safety Board, and placed the bag into a storage facility in Maitland, Maryland. Those same naturalists were responsible, some years earlier, for coining the term "snarge." Snarge is the avian tissue that's left over after a bird-strike (plate 5). It's a horrible word that onomatopoeically conveys the peculiarly destructive violence of acceleration in the Anthropocene.

What Is It? Snarge and the Snarge Matrix

If we extrapolate beyond the case of birds and planes, snarge can more helpfully be used as a way to characterize all collisions between forms of fossil-fuel based human mobility and solar-based animal mobility. Thus, snarge is the sum-total of all the dead or decaying tissue left over when plane hits bird, car hits mammal, boat hits manatee, ship hits whale, train hits cattle or elephant, and so on. It's difficult to calculate the biomass of snarge, though many have tried. We can say, with some degree of certainty, that in the United States alone between 1 and 2 million deer are snarged annually. They are among the very rough approximation of 365 million vertebrates—including twenty-one species that are threatened or endangered—who fall victim to automobile traffic in the United States. This annual number of snarged animals also includes 1.2 million dogs, 5.4 million cats, and between four and five thousand humans. Planes in the United States strike approximately 10,000 birds per year—a very conservative estimate. Between 10 and 20 whales are killed by marine traffic and between 80 and 100 manatees meet their end by Floridian propellers. Cumulatively, these organisms represent a collection of animals that occupy, as Robert Knutson put it in his field guide *Flattened Fauna*, "a habitat almost unique to the twentieth century. . . . Fast cars and hard-surfaced roads have produced the entire flattened fauna described here in less than an eye-blink of evolutionary time" (1987, 3-4).

Knutson perceptively put his finger on the issue of novel habitats. Creatures that are vulnerable to become snarge move, live, and die in anthropocenic environments that are largely designed to accelerate industrialized human beings and their goods. Highways and their right-of-ways have been designed for the speed and safety of automobile and truck traffic, but they are also edge habitats that bring animals out of the forest. Airports contain wide swathes of grassland that provide suitable browse and protection for birds, ducks, and geese. Dredged canals in Florida provide wonderful habitat for manatees. After heavy snows, Alaskan elk sometimes use cleared railroad tracks to move over the landscape. All of these environments are designed and managed for the safe acceleration of industrialized humans. Like so much of the Anthropocene, they are built environments, infrastructures, naturecultures, hybrid spaces, landscapes, designer ecosystems.

These "transit-ecological systems" have largely been designed by civil and traffic engineers. They are ruled by a paradigm that couples speed with safety—the protection of both efficiency and human bodies. The plants, trees, sedges, and grass used to landscape this infrastructure are aesthetic and functional decisions that support the paradigm. Nevertheless, they are, using the phrasing of Emma Marris's similarly named book, the "rambunctious gardeners" of modernity. They are joined by departments of transportation who actively maintain these habitats. They produce reports, studies, and guides—new forms of knowledge and expertise that govern the management of a significant portion of the globe's overall land use. What species of grass are best to plant? What kinds of pesticides and herbicides should be used? When to mow, how high to mow, invasives or—with seldom noticed irony—native species?

These moving parts—wildlife, infrastructural environments, the people and technologies that travel through them, and the people who design, study, and maintain them—cumulatively make up a kind of "snarge matrix." It would certainly be fair to call this world by more familiar terms—transportation network or infrastructure—but "snarge matrix" highlights the unintended but nevertheless designed death of animal lives. Most of us spend little time thinking about the snarge matrix, which is a missed opportunity. Snarge thinking helps us to understand the new relationships between humans and nonhuman animals cohabiting the Anthropocene. Thinking about snarge is unpleasant, and the pervasiveness of the problem tends to lead one to despair. But if we listen carefully, snarge can tells us something that may be helpful to figuring our way out of, or perhaps just coming to terms with, the Anthropocene.

What Does It Mean? Acceleration and Killing

At its essence, snarge represents the completely irreconcilable conflict between two different forms of mobility. One form we simply call "animobility," a helpful term used by sociologist Mike Michael that refers to motions that are fueled directly by sun and muscle, the results of millions of years of evolutionary history. The newer form of motion harnesses fossilized sunlight from millions of years ago and allows a rapid acceleration of industrialized human beings. In the blink of an evolutionary eye, this acceleration makes snarge possible and even inevitable.

Anthropocenic forces have dramatically transformed the lives of nonhuman animals. For our purpose here, snarging created a new form of killing that is without historical precedent and with few contemporary analogs. With some regrettable exceptions, snarge is a form of death that is almost always already an "accident." Like ocean "by-catch" or mowing over a family of corncrakes, death-by-vehicle is unintentional, a byproduct of acceleration. And it has become, more or less, an acceptable form of killing. That is, we do not hold the driver of a car

morally or legally culpable for accidentally killing a deer, and with good reason. A driver has no intention of killing. But it is still a kind of killing that calls
for a thoughtful response. Now, it would be wrong to imagine that life can exist
without participating in some forms of killing, but a thoughtful species, Donna
Haraway notes in *When Species Meet*, should consign itself to become thoughtful
killers. Snarge rarely registers on our moral radar, probably because the agent of
killing—mass acceleration—is structurally baked into almost everything we do.
This is a peculiar malady of the Anthropocene; by participating in an accelerated
mode of life, we have involuntarily become thoughtless killers of wildlife.

It is tempting to think about the ways that animals have responded to this
new form of killing. We know, for instance, that some moose rear their young
close to highway environments infrequently visited by predators that are presum-
ably scared off by the sound of traffic. We also know that some species of wildlife
accommodate their daily movements to highways. And snarging has selective
power with a corresponding evolutionary effect; we think that cliff swallows in
western Nebraska may have developed shorter wings to dodge traffic. These are
stories of animal agency, and there is virtue in knowing them, but such knowledge only gets us so far. Since the key tenet of anthropocenic thinking is that
industrialized humans have exercised earth-systems changing agency over the
past 200 years, any normative thinking needs to start and end with a deeper
understanding of human agency.

The dominant human response by both transportation and natural resource
agencies has been to *mitigate* the damage. Indeed, the field of "animal-
hazard mitigation"—a form of wildlife management that, again, is uniquely
anthropocenic—is big business. Similar to bounties placed on wolves and coyotes,
this form of management codes wildlife as a hazard to accelerated humans—an
accidental problem in need of mitigation. And so the prevailing paradigm has
been to harden the infrastructure, to make the snarge matrix impervious to animals, to clear infrastructure of wildlife that gets in the way. For instance, the
"zero tolerance policy" for geese around JFK, LaGuardia, and Newark International has led to a contract with the USDA's Wildlife Services that routinely
culls thousands of resident Canada geese—not, incidentally, the migratory fowl
that brought down Flight 1549—from the municipal landscape. There are infre-
quent moments when the discourse of mitigation flips, as with the case of certain
endangered species. When it was learned that motor traffic presented the largest
threat to the endangered golden panther, a highway was engineered—under the
mandate of the Environmental Species Act—to shield the interstate and mitigate
the damage to an endangered species. This was successful for animals around

I-75, but less helpful for the panthers navigating the sprawling road system of southwestern Florida.

Mitigation seems like a logical response to the snarge matrix, but it is a limited and partial response in two related ways. First, it does not grapple with the fundamentally irreconcilable conflict between animobility and the movements of accelerated humans. The discourse of mitigation almost always prioritizes movements accelerated by fossil fuels. Second, mitigation is steeped in a refusal to accept transit-ecological systems as hybrid landscapes. This is the dream of impermeability, which can be conceived of as the inverse of America's wilderness ideology. If our post-frontier desire for wilderness can be characterized by the dogged erasure of humans from the landscape, then the invention of the railroad cattleguard and the cowcatcher show exactly the opposite desire—an attempt to create an infrastructure that erases animals from the landscape. In this sense, William Cronon's important critique of wilderness as nonhuman must be matched with equal force by a critique of the right-of-way as only human.

Digging Our Way Out of the Snarge Matrix

If we accept that the snarge matrix calls for a moral response and that our response so far has been inadequate, then we might try to search for historical and contemporary solutions for living in or beyond the Anthropocene. Indeed, there are other stories, more humane stories, that snarge can tell us. One healthy response might be to apologize, as Barry Lopez does in his brief recounting of a car trip from Oregon to South Bend. *Apologia* is an unconventional travel narrative. Whenever he encounters roadkill, he stops to move the carcass deep into the right-of-way. "The ones you give some semblance of burial, to who you offer an apology, may have been like seers in a parallel culture. It is an act of respect, a technique of awareness"(1998, 3). Others have apologized through art, poetry, and activism. Some apologists find ways to literally mend the broken bodies and turtle shells that litter the right-of-way. For instance, wildlife rehabilitation clinics around the world attempt to administer care to the casualties of the Anthropocene. These are all important responses that speak to a sometimes conscious ambivalence or critique of an accelerated lifestyle. Living in the snarge matrix comes with guilt. Therapeutic wisdom to the contrary, guilt is a powerful human emotion and a necessary first step to healing the wounds of the Anthropocene. We must understand that getting in a car, plane, or train, that ordering a book from Amazon—all are destructive acts that create snarge. Wildlife deserves an apology.

But it's important to move beyond an apologetics of roadkill. It is possible to create and recreate infrastructure that is less deadly to wildlife. Taking its cue from European and Canadian landscape ecologists, the new school of "road ecology" is beginning to conceive of highway infrastructure as a part of the ecosystem dynamics of the wider landscape. From this light, a road can be a little like a river—a part of an ecosystem that has a barrier effect but can also be a mechanism for channeling the flow of energy, water, nutrients, and animals. A good example of this idea in practice is a fifty-mile stretch of US 93 in western Montana. After a long period of conflict pitting the Montana Department of Transportation against the Confederated Salish and Kootenai Tribes (CSKT) on the Flathead Indian Reservation, MDOT created a *permeable* highway with culverts, underpasses, and overpasses that balances the imperative for human acceleration with animobility. When landscape architects, environmentalists, wildlife biologists, corridor ecologists, community advocates, and transportation experts sit at the same table to redesign a road, something interesting can happen. Through a more *civil* civil engineering, the snarge matrix can be more thoughtfully redesigned. This, at first glance, seems like an expensive undertaking but advocates of permeability point to our crumbling infrastructure. If bridges and overpasses need to be rebuilt, they ask, why not rebuild them as corridors that reconnect previously fragmented habitats? Our transportation infrastructure is the greatest cause and symbol of life in the Anthropocene; working our way out of the Anthropocene requires a new infrastructure

Thoughtful design can lessen the destruction caused by accelerated humans, but there really is only one surefire solution to the problem of these conflicting mobilities—the deceleration of industrialized humans. Ivan Illich, the Austrian Catholic priest and social critic, made this argument in his 1974 *Energy and Equity*. Illich was less concerned with snarge than he was with the impoverishment of the Global South and the role that energy consumption played in perpetuating inequality. All societies can benefit, he argued, by increasing their consumption (their speed) up to a discrete point. Beyond that point, social relations degrade and power and capital pool into the hands of a small group of technocrats. In an argument that was prescient of Pope Francis's recent encyclical, Illich called for the industrial world to slow down, to wean itself from an increasing consumption of energy in the interests of global equity. The top speed of a modern bicycle would suffice. Of course, the opposite has happened. Since Illich's time, the portion of humanity enjoying the fruits of acceleration has increased in number as developing nations adopt the mobility patterns of overdeveloped peoples. The virtue of slowness seems like an impossible dream.

But there are stories that highlight the possibilities of a decelerated world. The creation of no-wake zones in South Florida's canals and lakes provides some respite for the glacially slow movement of the Florida manatee. And an energetic lobbying group of scientists and conservationists has been able to decelerate the pace of global shipping traffic. In the interest of the endangered North Atlantic right whale, they have convinced the International Marine Organization to shift shipping lanes away from critical right whale habitat and, even more surprising, they have created seasonal speed limits for large ships that have no choice but to share their sea lanes with whales in what some conservationists call "the urban ocean." Even more inspiring stories come from outside the United States. D. Rajkumar of the Wildlife Conservation Foundation has recently been able to completely halt night traffic along several highways that move through the Bandipur Tiger Reserve outside of Mysore, India. And the UK-based nonprofit Froglife mobilizes kids to safely slow down car traffic that threaten amphibians and reptiles crossing British roads.

These are small but significant countercurrents. And on some level, they have much in common with other environmental movements currently in fashion: the slow food movement, initiatives for walkable and sustainable cities, car-sharing, and the down-shift movement, to name a few. Indeed, some travel authorities believe that we have hit the point of "peak travel." Accelerated humans may be growing travel-weary and we may have saturated our desires and technologies of acceleration. Even more, the "Complete Streets Coalition" has highlighted the problem of hugely disproportionate pedestrian deaths and injuries among people of color in many American cities. They are reframing the issue of sustainability as a movement of social justice. In a cruel irony, a movement that began with a bus boycott continues to play out, but this time we are denying public transportation to our most at-risk citizens so they can fend for themselves on the streets, in a not-too-distant future, filled with driverless cars. The critique of mass acceleration is becoming *de rigueure*.

But we knew this already. How does thinking about snarge add to the discussion of the Anthropocene? And aren't there more important things to worry about? Of course. But thinking about snarge has three particular virtues. First, it aims our focus at the root cause of anthropocenic climate change—an accelerated mode of existence that is impossible without fossil fuels. Indeed, I might amend Aldo Leopold's land ethic. "A thing is right when it tends to preserve the integrity, stability and beauty of the biotic community. It is wrong," Illich and I would argue, when it moves faster than the speed of a bicycle (Leopold 1987, 224–25). If we found ways to eliminate inefficient acceleration, we would lessen the snarged victims of the world, but there would also be some obvious beneficial cascading effects. Second, snarge is a reminder of how quickly industrialized humans accommodate to, and normalize, anthropocenic change. Roadkill, like pedestrian deaths, were a source of moral outrage in the 1920s when automobiles began roaming the industrialized world en masse. But then we just got used to it. Roadkill became an acceptable—no matter how regrettable—casualty of modern life. Snarge needs to be an uncomfortable reminder of that shifting baseline, an indictment not only of acceleration, but of the very process of becoming comfortable with structural violence. Finally, thinking about snarge is a powerful reminder that despite the many intellectual countermovements to animalize *Homo sapiens*—evolution, the id, sociobiology, deep ecology, biocentrism, ecocentrism, and so on—accelerated humans wield a terrifying power that distinguishes us from the experiments of the living.

The Anthropocene is less a geological epoch than it is a story, and that is why journalists and scholars in the humanities and the social sciences have been so eager to embrace—or at least tangle with—the concept. Thinking about snarge leads to experimenting with new stories—stories about the virtues of slowness, of moving our bodies and our goods in a more just and humane rhythm. The moral of snarge's story is not, to be clear, about the virtue of stasis. Rather, it is just another iteration of the stories of Icarus, Prometheus, Frankenstein, Dr. Strangelove, the Tortoise and the Hare. Think about it this way. There are those among us who will go out of their way to run over a tortoise crossing a road, and there are others who swerve out of the way to avoid the tortoise, and there are those who get out of their cars to help the tortoise to safety. These are all anthropocenic responses. Snarge would have us selling our cars and taking the bus.

BIBLIOGRAPHY

Haraway, D. 2007. *When Species Meet*. Minneapolis: University of Minnesota Press.

Illich, I.1974. *Energy and Equity*. London: Calder and Boyars.

Knutson, R. M. 1987. *Flattened Fauna: A Field Guide to Common Animals of Roads, Streets, and Highways*. Berkeley, CA: Ten Speed Press.

Leopold, A. 1997. *A Sand County Almanac and Sketches Here and There*. Reprint, New York: Oxford University Press.

Lopez, B. 1998. *Apologia*. Athens: University of Georgia Press.

Marris, E. 2011. *Rambunctious Garden: Saving Nature in a Post-Wild World*. New York: Bloomsbury.

Marine Animal Satellite Tags

Nils Hanwahr

37°43'51.0"N 123°05'18.0"W—Pacific Ocean,

Off the Farallon Islands

The team of researchers has for a while been hurling scrambled fish overboard to attract sharks. They are at one of the most famous white shark hotspots in the world: the Farallon Islands off the coast of San Francisco. It is a familiar image from wildlife documentaries: the shark is lured in and then turned on its back so that it faints; while it is restrained at the side of the boat, suntanned marine biologists attach a device to its fin; the researchers record the size of the creature before turning it over again and releasing it. Appearing to recover swiftly from its disoriented state, the fish flicks its tail once or twice and dives beyond the depth breached by sunlight and human sight. But now, as the shark continues on its invisible paths in the ocean, it is leashed to an invisible tether that reports and tracks its every move.

Satellite tags are attached by marine biologists to many kinds of large animals living in the oceans, including sharks, whales, and sea elephants (plate 6). The tags collect information about each animal's position, speed, diving patterns, and marine environment through which it swims. The collected data is sent via satellite back to the researchers on land who can use it to study the behavior of sea creatures.

Tagging a marine animal with a high-tech device endows the creature with a kind of agency that could only arise in the Anthropocene—the satellite tag records the perspective of a creature that has existed and evolved on the planet for a much longer time than have humans. Agency only registers on our human scale by leaving a trace, and in the twenty-first century that means registering life forms and environments as digital data. We incorporate remote environments into our digital representations of nature. In doing so, we "keep projecting ourselves into landscapes we're not equipped to cross in the flesh," as science writer Diane Ackerman puts it (2014, 173). Yet, in expanding our cognition via cyborg creatures, we encounter human impacts on biospheres that have hardly registered as part of our environmental imagination.

Technology such as the satellite tag enables humans to become fellow travelers with other creatures and shift the boundaries of an "ethics of proximity" in both space and time that fuses the ecological and digital entanglements of our Anthropocene planet. The proximity of creatures in human lives has been a basis of thought in animal and environmental ethics, yet the scales of the lives of creatures such as a shark continue to challenge our moral imagination. It is here at the Farallon Islands, in proximity to Silicon Valley, that the shark we are about to follow has been integrated into an Anthropocene infrastructure no less complex than the worldwide web of smartphones that have become our daily companions and watchdogs—the same smartphones that enable the general public to follow on Twitter, for example, the movements of satellite-tagged animals.

In following a tagged shark on its path, one can trace the story of the satellite tag technology as part of the history of what historian of technology Paul Edwards calls the "vast machine" of our global scientific infrastructure. This technology challenges our notions of the high seas as a free and endless space. New ways of visualizing marine animal movements via online platforms turn mere representation of ocean space into a lived space in an "internet of animals" (Pschera 2014). The sublime ocean that amazes and overwhelms the human mind can now be viewed through the frame of some of its oldest inhabitants.

Knowledge gained through tagging marine animals is not only challenging notions of space, but also enables new ways of dynamic ocean management that take the agency of its inhabitants into account, thus shaping the concept of the existing law of the sea, and new kinds of *symbiopolitics*, a term that describes "the governance of relations among entangled living things" (Helmreich 2009, 15). Such a politics is called for in any critical awareness of the Anthropocene. And mapping the impact we have on the planet also calls for going beyond land-based habitats, bringing visibility to the manifold ways we are encroaching on biospheres that we are not equipped to cross in the flesh. In enlisting life forms to explore realms we will never visit, we find that other human influences have arrived there before us. We've gotten ahead of ourselves—welcome back to the Anthropocene!

32°48'13.2"N 117°26'30.4"W—Pacific Ocean, Off La Jolla

The tagged shark emerges next off the coast of La Jolla, San Diego, the site of the world-famous Scripps Institute of Oceanography. Scripps has been monitoring the oceans and its creatures for science as well as for their immediate neighbors, the US Navy, which has several huge aircraft carriers stationed in this important military harbor on the US Pacific Coast. We also find Sea World, the infamous marine theme park where the first experiments in tracking large marine animals, in this case orcas, were carried out. Starting with radio telemetry, Sea World enabled scientists to take the first steps in deploying a technology that some regard with the same suspicion as they regard marine animals being held in captivity.

Tagged animals help to incorporate remote environments into our digital representation of nature that we perceive as mediated by virtual data collections. However, the technology that enables data gathering at such a large scale and in near real-time is comparatively new. Also, the pop-up satellite tags rely on an infrastructure of satellites that requires even more sophisticated technology and significant investments. Ocean monitoring systems as well as satellite-based communication systems have their origins in military applications and research of the Cold War.

Navy experiments with marine mammals and sharks had been carried out during the Cold War, utilizing and training marine animals for military purposes. As Diane Ackerman describes: "plans included remote-controlled sharks (with electrodes in their brains) designed to sniff out bombs and explosives. . . . For instance 'Mk 4 Mod 0' is a dolphin trained to detect a mine near the seabed and then attach an explosive charge to it; 'Mk 5 Mod 1' is a sea lion used to retrieve mines during practice maneuvers" (2014, 145).

Early sound- and buoy-based monitoring networks were deployed by Navy oceanographers to monitor for nuclear tests by rival powers and track the movements of Soviet submarines in the Atlantic and Pacific oceans (Oreskes and Krige 2014). As research by historian of science Naomi Oreskes shows, oceanographers had been involved with military purpose-built infrastructures all throughout the Cold War. After the collapse of the Eastern Bloc, the US Navy declassified the maps it had kept of the sea floor, which also marked a turn for the oceanographers' occupation. Like many military technologies, new applications were found by those biologists and oceanographers who had previously worked for the US Navy and were now starting to apply their skills to research.

The first satellite-based tags were built in the 1990s and were manufactured by a few small workshops, located mostly on the East and West coasts of the United States. For example, one company called Desert Star started out as a defense contractor building homing devices as safety technology for Navy divers. Expertise for locating military divers underwater led to the development of satellite-based tags that are now sold to research institutions worldwide.

Another major technology that oceanographers use in their work today are remote-operated vehicles, called ROVs, tethered robots equipped with cameras and mechanical arms that can collect samples from the sea floor or search the sea floor with lights and cameras. Stefan Helmreich, an anthropologist, describes the use of ROVs by scientists as a prosthetic extension of human capabilities, enabled by a "submarine cyborg" that is intimately connected to its human operator (2009).

In the case of animals tagged with satellite technology, another level of interconnectedness is built in since, as is the case with ROVs, humans extend their cognitive capabilities to the submerged world. In addition, the prosthetics that expand the cognition of human researchers also expand the cyborg capabilities of the creature. Correspondingly, we watch and surveil creatures not just in Sea World's aquariums, but via apps and interfaces through the "portholes" of our smartphone screens. Sea World's practice of keeping large marine mammals in captivity has been justifiably criticized. Similarly, attaching tags by barbed darts may cause pain to the animal as well as impede its swimming. Also, one might wonder if turning an animal into a data point does not in itself entail an act of violent reduction into a digital infrastructure.

22°33'30.0"N 134°49'54.8"W—Pacific Ocean, White Shark Café

The creature carrying a satellite tag, however, is never fully tamed by our research infrastructure. It will always have abilities that surpass our own, roaming spaces that we cannot reach, and perceiving what we cannot sense, such as the shark's ability to perceive electrical stimuli. Thus, being connected to our data networks, the animal is on "a two-way tether" (Ackerman 2014).

We next meet the shark out on the open sea, halfway between the US Pacific Coast and Hawaii, a space that is rarely considered by most terrestrials. Yet, for the animal the offshore constitutes a different kind of space, it is its natural habitat. Among scientists, this congregation area, discovered through extensive tagging studies, is known as White Shark Café. Tag technology helps to map our impact on the environment, renders a "scientific" agency to the animals involved,

Figure 5. The global tagging of a pelagic predator incorporated into Google Earth.

projects our human agency into remote landscapes, and creates novel geographies and locales such as the White Shark Café. This creates the potential for the scales of ocean life to register in a human context, although the difference in power remains clear. It is the sharks that remain endangered by us humans; after all, there is no opt-out option for a tagged shark living in an ocean under the influence of anthropogenic climate change.

However, by tagging an animal we do not attempt to train or tame it, we merely use technology to trace an entanglement that already existed, an entanglement that no creature of the Anthropocene can escape. Thus, the "cyborg" research technology enables our global cognition systems to come "full circle," as the mission statement of the research program Global Tagging of Pelagic Predators (GTOPP) puts it:

> Our objective is to understand the factors that influence animal behavior in the blue ocean and to build the tools required for protecting their future. In addition, animals carrying tags quickly become animal ocean sensors and can contribute millions of data records that can help climate scientists build a better understanding of planet Earth.[1]

The resulting data is then incorporated into tools such as Google Earth (see fig. 5). Thus, we are able to view data from individual animals very much like we would view data from commercial airline flights when tracking our journey to a holiday destination.

Yet tagged sharks, like humans, are not necessarily willing subjects in the surveillance infrastructures in which both human and nonhuman creatures now find themselves enmeshed. It is one critique leveled among those denouncing the current deployment of tens of thousands of tags in the oceans. As with any experimental system, observation is likely to change the behavior of the observed creature. Dragging a satellite tag that is attached to skin by a barbed dart may well impede an animal's movements. And while researchers point out that a shark's fin, to which the tags are attached, does not contain nociceptors, which sense pain, the fin can potentially be damaged and become infected. Remnants of the attachment device can also attract parasites. In addition, a review study of tagging has recommended that colorful tagging devices be avoided, since they make the shark more visible to its prey and can cause changes in behavior. Despite such ethical objections, scientists defend animal tagging by insisting that some pain to the individual animal is preferable to not knowing more about the ways in which environmental change is impacting the oceans as a whole (Hammerschlag et al. 2011). Yet it may be our ideas of the "ocean as a whole" that the painful transformation of animals into ocean sensors can help us rethink.

21°12'37.6"N 157°57'50.1"W—Pacific Ocean—Off of O'ahu

Hawaii, the next stop of our shark, is the most western outpost of the United States and the site from which the country projects its power over vast distances in the Pacific Ocean. President Barack Obama announced in 2012 that the United States would take the Pacific region as its new strategic focus. Despite their apparent barren emptiness, oceans are strategic spaces where politics and power shape environmental and social relations enacted over large geographical distances.

The oceans as spaces of power and territory have changed over the centuries. The ancient Roman looked no further than the Mediterranean Sea, which they called *mare nostrum*. With improved technology and the rise of sea trade, oceans served as the transportation surfaces of imperial and economic ambitions. The Dutch lawyer Hugo Grotius developed the seventeenth-century notion of *mare liberum* in response to the Dutch and British scramble for South Asian resources. The freedom of the seas was to be maintained as a neutral space of international power and trade relations. A wider public interest in the oceans beyond their role as transport surfaces, both in popular science and as a tourist destination, is actually fairly recent, with the seaside turning into a holiday destination only in the nine-teenth century.

Today, most popular accounts, even those problematizing the limits of the ocean's resources as an environment, present the high seas as a vast endless space of sublime depths and celebrate the freedom of the seas. These notions make it harder for laypeople and policymakers to grasp oceans as a space that can actually be impacted by human activity. We find it easier to imagine mining for resources on the moon than to police offshore spaces effectively or begin to deal with the great Pacific garbage patch. After all, how could an "endless" ocean have a limited capacity for pollution or fishing?

Research based on animal satellite tagging has contributed to readjusting the perception of the oceans and the limits to their exploitation. Also, the effects of climate change are manifesting in vastly dispersed locations while our influence on the scale of the global climate system within only a few human generations appears to speed up geological time. The sheer geographical size of the oceans spans many spheres of political influence. Our human impact on ocean life is hard to fathom. Satellite tagging can add another layer of ocean life that has a chance of being taken into account when human actors attempt to regulate and police ocean spaces.

22°59'00.2"N 108°20'15.8"W—Pacific Ocean, Entering Baja California

Policy is a way of encasing and formalizing our environmental imaginations and can, at times, be disconnected from its framing of issues and spaces. The shark has, unnoticed by state authorities, crossed the border between the United States and Mexico. The satellite tag keeps on tracking and gathering data irrespective of political borders. While many Mexicans attempt to cross the US border every day to seek out a more secure and prosperous life in the north, marine animals make annual migrations south to one of the major marine nurseries on the planet, Baja California. By tracking marine animals' movements or observing migrations such as those of the grey whales, we come to realize that human borders are not nec-essarily obstacles to animal movement even though they may impede effective action and policy in preserving the animals' natural habitats. The creature does not respect political borders nor is it concerned with the challenge of coordinat-ing international policy initiatives.

The United Nations Convention on the Law of the Sea (UNCLOS) was conceived in 1984, yet has to this day not been ratified by some major nations, notably the United States. When President Ronald Reagan rejected the UNCLOS, he cited American commercial and military interests. Basically, UNCLOS commits all oceans inside a twelve-nautical-mile coastal zone to the administration of respective states. It also gives states the right to exploit resources on the continental shelf within an exclusive economic zone (EEZ) of 200 nautical miles, which has been expanded by some countries to 350 miles offshore.

Outside of the EEZs of nation-states, the high seas are governed by the rules of UNCLOS and by agreements such as the convention on offshore dumping and various fisheries laws. However, there is no international body that effectively enforces or polices this zone. Efforts to establish such a body have been widely rejected with reference to the concept of *mare liberum*, the freedom of the seas.

Obviously, ocean creatures do not adhere to the arbitrary boundaries of EEZs or territorial waters. Thus, these ocean dwellers are difficult to handle within a framework wherein the notion of freedom of the seas views oceans as empty space to be traversed by unimpeded shipping. Consider a standard world map of the type often found in classrooms. Physical features such as mountain ranges and rivers and political features like national borders are included in the map. But the oceans are mostly presented as a large empty space in light blue. Oceans are merely a surface to be traversed and to be kept clear of obstacles, such as piracy or nationalistic restrictions.

How will we uphold such a view as we come to know so much more about how other nonhuman actors use this same ocean space, traversing it for purposes entirely different than mercantile shipping? The consequence of such knowledge might be to close certain areas to fisheries to protect migratory species. This is the idea of dynamic ocean management (DOM), which proposes to use information collected by tagging, biologging, and remote sensing to adjust protected areas to accommodate animal migrations and other behaviors. While such interventions are technologically possible, much work remains to be done in policy and enforcement: "fishers could potentially challenge those organizations that collect and process the data upon which decisions are made" and so to "use DOM effectively, the science and technology required for DOM needs to be better integrated with the enabling policy" (Hobday and Maxwell 2014, 126, 154). How this integration of technology and policy is put into practice remains to be seen. Tracking technologies could just as easily be used to fish more rather than less.

0°04'13.4"S 90°16'00.1"W—Pacific Ocean, Off of Galapagos Islands

Swimming out of Baja California, the shark has turned south and arrives at the cliffs of the rough Galapagos Islands. The shark circles the unique ecosystem of the Galapagos Islands that, by virtue of their insularity, enabled Darwin to study natural selection. Yet, the shark remains tethered to scientists on the Californian coast who have turned it into a cyborg so as to better understand and study its wild nature. What the scientists have found: invisible boundaries, global digital networks of information and commerce, and military and financial power projected on vast scales. The Anthropocene reminds us of how borders, including species borders, political borders, and geographical borders, are malleable: they can be opened and closed by us.

"Anthropos" means human, but the meaning of "cene" is multifaceted. "Cene" is usually taken to mean era. However, "Holocene," the name of our current (or immediately previous) geological era, means "entirely recent time." So "cene" can also mean recent or now. Thus, I would like to think of the Anthropocene as the "recent-human-now." The true novelty about the Anthropocene is not that human activity has altered the face of the planet, but rather the speed at which this has happened. The Anthropocene takes place now in front of our eyes and because of recent human activity. The fact that we find human influence everywhere, that we find ourselves everywhere, exceeds our individual grasp. The speed at which we transform the world that we are part of also still exceeds the capacity of science and technology to actually monitor this impact.

It is this entanglement that we increasingly manifest in our notions of ecology and in our digital infrastructures, since "the digital network, indeed, is the counterpart and in some sense the master trope for the ecological connectivity with which it fuses at the end" (Heise 2008, 209). Satellite tags allow us to accompany fellow creatures in the time of now, to observe our recent human impact; this technology enables us to look at ourselves through the lives of other entities that roam this earth, and even contemplate ways of life in ocean environments that we, nevertheless, may never comprehend. Yet, reflecting on what it means to track a wild creature as an integrated cog of a global surveillance system can, at the same time, raise questions about our utopias of all-encompassing digital networks that envelop the life-worlds of individual humans and our fellow creatures in ever more complex infrastructures of control.

NOTE

1   Tagging of Pelagic Predators, http://gtopp.org/about-gtopp/programs/background.html.

BIBLIOGRAPHY

Ackerman, D. 2014. *The Human Age. A World Shaped by Us*. New York: W. W. Norton.

Hammerschlag, N., et al. 2011. "A Review of Shark Satellite Tagging Studies." *Journal of Experimental Marine Biology and Ecology* 398:1–8.

Heise, U. K. 2008. *Sense of Place and Sense of Planet: The Environmental Imagination of the Global*. New York: Oxford University Press.

Helmreich, S. 2009. *Alien Ocean: Anthropological Voyages in a Microbial Sea*. Berkeley: University of California Press.

Hobday, A. J., and S. M. Maxwell. 2014. "Dynamic Ocean Management: Integrating Scientific and Technological Capacity with Law, Policy, and Management." *Stanford Environmental Law Journal* 33:125–65.

Oreskes, N., and J. Krige, eds. 2014. *Science and Technology in the Global Cold War*. Cambridge, MA: MIT Press.

Pschera, A. 2014. *Das Internet der Tiere. Der neue Dialog zwischen Mensch und Natur*. Berlin: Matthes and Seitz.

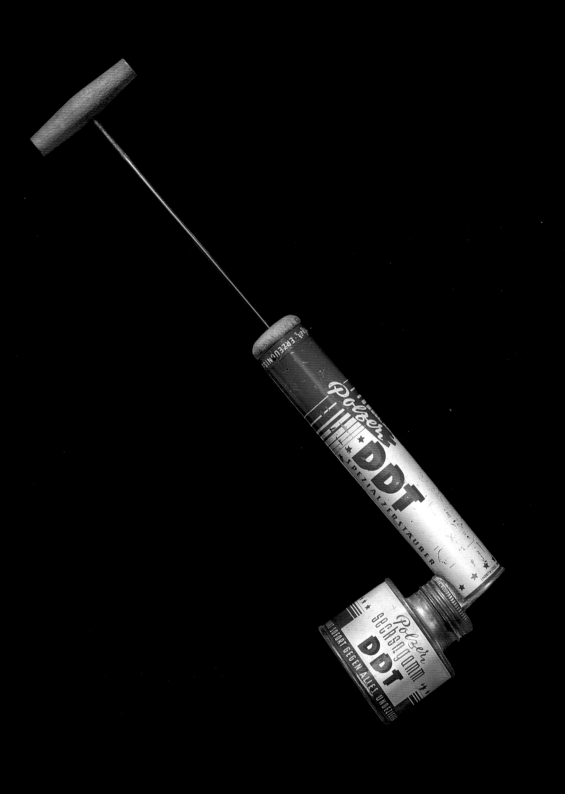

20

ENGINE #1

BYPASS OGV PLATFORM
OUTER FLOWPATH

3 'O'CLOCK ALF
1-29-09
US Air 15??

Isotope
Study

Dolbear

1 of 6

Friends Meeting house

corner of Arch

Plate 1. The manual pesticide spray pump. Photo: Tim Flach.

Plate 2. A jar of sand. Photo: Tim Flach.

Plate 3. Concretes. Photo: Tim Flach.

Plate 4. Huia Echoes. Photo: Tim Flach.

Plate 5. Snarge. Photo: Tim Flach.

Plate 6. Marine animal satellite tags. Photo: Tim Flach.

Plate 7. Artificial coral reef. Coral Morphologic (Colin Foord and Jared McKay), 2011. Location: Key Largo, Florida. Depiction: Staghorn coral (Acropora cervicornis) on a PVC substrate, grown by the Coral Restoration Foundation.

Plate 8. Cryogenic freezer box. Photo: Tim Flach.

Plate 9. The monkey wrench. Photo: Tim Flach.

Plate 10. The Germantown calico quilt. Photo: Tim Tiebout. Courtesy of Laura Keim, Curator, Stenton, Philadelphia, Pennsylvania.

Plate 11. The mirror. Photo: Tim Flach.

Plate 12. Objects from Anna Schwartz's Cabinet of Curiosities. Artist: Judit Hersko. Credit: Tim Flach.

Plate 13. Technofossil. Maker: Jared Farmer. Photo: Tim Flach.

Plates 14a and 14b. *Davies Creek Road*. Painting by Trisha Carroll and Mandy Martin. Photo: Tim Flach.

Plate 15. *Plowshares* film. Photo: Tim Flach.

Artificial Coral Reef

Josh Wodak

The Acknowledgment of the Anthropocene

Two gentlemen sit facing each other in the Oval Office. It is May 8, 2015, a date undistinguished except that it is the eighty-ninth birthday of the more elderly of the two gentleman: Sir David Attenborough. He is meeting President Barack Obama for the first time. The world's most powerful person is hosting the world's most influential naturalist for a wide-ranging discussion about "the future of the planet, their passion for nature and what can be done to protect it."[1]

Obama, having confessed his starry-eyed admiration for a man whose nature documentaries inspired his boyhood explorations of coral reefs in Hawaii, asks what Attenborough's favorite experience and place in the world is. He answers that it is the sensation of "the moment you first dive on a barrier, on a coral reef . . . with this multitude of multi-coloured organisms, the like of which you've never seen before." Obama nods and smiles in silent recognition of his own sensations of being among coral reefs.

Having established their mutual admiration for nature in general, and coral reefs in particular, the men turn their discussion to their respective perspectives on the challenges facing coral reefs, given that one-third of marine species depend on such reefs even though they constitute only 0.1 percent of ocean surface area. Attenborough has just returned from the Great Barrier Reef in Australia, the biggest coral formation and largest living structure on the planet, and a place he has returned to over more than sixty years. Equipped with his training in geology and zoology, he is able to anecdotally recount the more visible changes he has witnessed over time: namely, the rapid increase of coastal population and associated increases in the discharge of industrial effluents onto the reef. Attenborough is also able to contextualize these local-scale problems in their global historical context: his first visits to the Great Barrier Reef in 1950 coincided with a dramatic spike in the Great Acceleration, when human activity began to impact Earth system processes at rapidly increasing rates, visible in graphs depicting escalating greenhouse gas levels, ocean acidification, deforestation, and biodiversity loss.

The exchange between Obama and Attenborough reveals the crux of why coral is the canary in the coal mine of climate change, and coral's resonance within a Cabinet of Curiosities for the Anthropocene:

ATTENBOROUGH: The real problem on the reef is the global one, which is what is happening with the increase in acidification and the rise in the ocean temperature and the Australians have done research on coral now and they know for sure it will kill coral.

OBAMA: Right.

ATTENBOROUGH: It will kill the species of coral and what they're concerned about now, is, I mean, that seems almost inevitable. What it seems now is can they, can they find the right species to maintain the reef's population?

OBAMA: Right, so really there's a mitigation strategy that they're trying to come up with but what we're seeing is global trends that depend on the entire world working together.

The stern warnings Attenborough refers to are from major scientific organizations researching the volatile tolerance thresholds for marine organisms to survive anthropogenic ocean warming and ocean acidification. The urgency and severity of these warnings have given rise to converging aspirations between conservation biology and environmental engineering. That is, the "mitigation strategy" Obama refers to is in response to how anthropogenic climate change constitutes an unplanned and unintentional experiment with the biosphere and atmosphere. The "strategy," then, is Attenborough's hope that scientists can "find the right species to maintain the reef's population." This act of "finding" is closer to an act of (re)making nature: from the accident-of-the-Anthropocene to human-designed "nature"; both are expressions of intentional influences exerted by humans. Such "finding" encompasses the euphemism of "Assisted Evolution," involving genetic modification to increase the tolerance threshold of coral against projected rates of acidification and warming (van Oppen et al. 2015, 2307–13).

The notion of (re)making nature finds its foremost expression in two related forms of environmental engineering that coalesce over coral: synthetic biology and geoengineering. While the former operates at the micro-scale of life itself and the latter at the macro-scale of planetary biosphere and atmosphere, they share a core concern: to intentionally remediate biophysical environments through their "design" of life and climates respectively. These fields, where cutting-edge science meets an engineering mindset, have been the focus of reports by Britain's Royal Society, which defines synthetic biology as "the design and construction of novel artificial biological pathways, organisms or devices, or the redesign of existing natural biological systems" (2009b) and geoengineering as "the deliberate large-scale manipulation of the planetary environment to counteract anthropogenic climate change" (2009a).

These micro- and macro-scale designs converge in the act of (re)making nature through artificial coral reefs. Coral reefs already problematize the boundaries between the living and dead and between animal, plant, and mineral: they are unique in simultaneously occupying all these categories and properties in such a way that cannot be disaggregated from one another. The animal element of living corals secretes a mineral substrate to form continuously growing structures that are technically dead. The colonies of small animals that collectively constitute a coral reef enjoy a symbiotic relationship with zooxanthellae, the microscopic algae they live among, which provide the constituents of a coral reef the majority of their nutrients. Due to such bewildering symbioses between life and death and inclusion of creatures across taxonomic kingdoms, coral reefs have been highly coveted in historical cabinets of curiosities.

An artificial coral reef sample for the Cabinet of Curiosity draws upon this historical sense of wonder and focuses our reverence on a tangible encounter and an ethical quagmire: designing nature as a means of conservation. The artifact is based on the work of the Coral Restoration Foundation, a Florida-based nongovernmental organization that grows colonies of staghorn coral in nurseries and then seeds them onto concrete and metallic structures in the open ocean, as both literal and metaphorical life support (plate 7). The planned object in the Cabinet is a segment of nursery-propagated staghorn coral embedded in a bedrock of PVC (polyvinyl chloride) pipes, preserved in formaldehyde, and presented in a glass container.

The size of the container, measuring one cubic foot, refers to the Biocubes of the same size that scientist David Liittschwager used for his project titled *Life in One Cubic Foot*. For his project, a team of scientists placed hollow green-framed cubes in different biomes around the planet to measure the biodiversity of the life forms that passed through the cube during a twenty-four-hour period. The relatively small size of this Cabinet display provides a metaphorical magnifying glass through which to view the planetary-scale processes, issues, and consequences that artificial coral reefs evoke, much like the way a glass container magnifies its interior objects through refraction. No longer ambiguously existing between life and death, this preserved specimen is seemingly frozen in time. Its symbiosis between algae, secreted minerals, and coral polyps is augmented by the human-nature hybridity of PVC, a literal life-support structure, a surrogate substrate for a natural reef. As a biological specimen, it is rendered presentable in a museum by replacing its natural surroundings of water with a toxic tissue fixative and embalming agent, formaldehyde.

To wade further into the ethical quagmire of how environmental engineering coalesces over coral reefs, we must turn first to coral itself, to see how it is a lens through which we can simultaneously focus on micro- and macro-scales, to view the Anthropocene writ large and small, to reconstruct the past and to forecast the future, and, in so doing, to accept the Anthropocene as historical accident meeting contemporary intention.

The Advent of the Anthropocene

Through May and into June [1770], *Endeavour* sailed north, arcing northwest, following the Great Barrier Reef coastline. On the evening of June 10, when most of the men were sleeping, the ship struck coral, stuck fast, and began leaking. Quick thinking and decisive action by [Captain] Cook and his men—pumping furiously and jettisoning fifty tons of decayed stores, stone ballast, and cannons—kept the ship afloat and allowed a temporary underwater repair. (Delaney 2010)

The transit of the planet Venus between 1761 and 1769 marked the impetus for the first genuinely worldwide scientific collaboration. The quest to analyze this rare celestial event required nations to work together for a common purpose—even those otherwise working against one another, such as Britain and France who were at the time at war with one another. Scientists stationed themselves in dozens of locations across the planet to take measurements of Venus passing between the earth and the sun all over the world. Through devices emblematic of the Enlightenment, including telescopes and quadrants, their observations of the transit were subsequently combined to yield the first accurate measurement of the size of the solar system. This transit of Venus also occurred at the outset of the industrial revolution, now regarded as the advent of the Anthropocene.

This worldwide scientific collaboration led to one of the first major scientific encounters with coral reefs, in particular the Great Barrier Reef. Captain Cook had been commissioned by the Royal Society to travel to the South Pacific on the *Endeavour* to undertake observations of the transit of Venus. On board was Joseph Banks, renowned British explorer and naturalist who later became president of the Royal Society; Banks researched coral throughout the expedition.

Following Cook's observation from Tahiti in 1769 of the transit of Venus, his ship ran aground on the Great Barrier Reef in June 1770. As historian John Delaney recounts in the above quotation, during the ship's emergency fifty tons of materials and equipment, including metallic objects such as cannons, were jettisoned onto the reef below. Lying littered across ocean sea floors, the remnants of shipwrecks form accidental artificial reefs as coral use the properties of various metallic structures as scaffolding on which to grow reefs. Similarly, in artificial reefs, coral takes advantage of human-made substrates where the relationship between synthetic biology and geoengineering comes more sharply into focus.

Coral offers another means of determining human influences on the biophysical environment. Reefs, which can live for thousands of years, present paleoclimatological records through the year-by-year accretions left by coral that forms the reef itself. Due to coral's high sensitivity to ocean warming and ocean acidification, reefs contain precise records of changes in past climates which palaeontologists use to analyze prior extinctions. For instance, the living structure of the Great Barrier Reef is between 6,000 and 8,000 years old and emerged over the Holocene, while structures from prior geological ages are now submerged and fossilizing.

Through their fidelity of registering miniscule environmental changes, coral reefs also present the visible effects of human activity as a geomorphic force. Six months after Attenborough and Obama met, these effects became all the more apparent when the Great Barrier Reef underwent its worst coral bleaching in recorded history during the 2015–2016 Southern Hemisphere summer, with 93 percent of the coral detrimentally affected. Justin Marshall, chief investigator of the citizen science program Coral Watch, undertook a survey another six months later, during the Southern Hemisphere 2016 winter, when cooler winter waters would reveal the ongoing bleaching effects on the coral. His conclusion that sections were already in "complete ecosystem collapse" has furthered interest in the contentious notions of designing aquatic life forms that may be able to exist in conditions rendered inhospitable to existing nature (Slezak 2016).

Coral: Bridging Earth and Venus

During the most recent transit of Venus, on June 5–6, 2012, another genuinely worldwide collaboration was taking place, although of an artistic rather than scientific nature. Australian artist Lynette Walworth was staging *Coral: Rekindling Venus* at planetariums around the world, including the film's world premiere at the American Museum of Natural History. Her installation offered an inverted encounter with coral reefs: film of reefs was projected on the ceiling above viewers' heads, so that viewers looked up rather than down at reefs on the sea floor. The title and the timing of Walworth's art installation references what had transpired in the centuries since Cook and others around the globe observed Venus passing directly between the sun and the earth. By 2012 on Earth, coral had become something artificially cultivated, cloned, and engineered, including being encouraged to grow on deliberately created artificial reefs deployed around the world. The substrates for these artificial reefs include diverse military-industrial detritus such as train carriages, tanks, and car tires, as well as modular structures made of plastics, Fiberglas, concrete, and hydraulic cement, and even metallic scaffolding through which low-level electric current is run to cause limestone buildup on the metal and molds.

To focus our gaze on the planetary-wide implications of coral, Walworth's film premiered during the 2012 transit of Venus harking back to the international collaboration required to analyze the 1761 and 1769 transits of Venus. In this way, Walworth points to the cross-sector collaboration required to face the existential challenges posed by the Anthropocene. This connection between coral reefs and Venus is also made in the American Museum of Natural History where an exhibit in the Gottesman Hall of Planet Earth draws an Earth–Venus connection via coral that is orders of magnitude more catastrophic than the worst recorded global coral reef bleaching of 2015–2016, more catastrophic even than scientific models that forecast the loss of the majority of coral reefs by the end of the twenty-first century. The exhibit draws a connection to the Venus Syndrome: the most widely supported theory for how Venus lost its atmosphere and became the scalding, uninhabitable planet it is today. Dr. James Hansen, originally an atmospheric physicist and then director of the NASA Goddard Institute for Space Studies for more than thirty years, is the foremost proponent of the Venus Syndrome. Hansen is one of the world's preeminent climatologists and one of the most public voices to argue that the Anthropocene does not represent "just" the sixth major extinction since the Cambrian explosion 570 million years ago, but instead that the long-term trajectory of runaway climate change would lead to the eventual loss of the atmosphere on Earth, in a manner similar to what occurred on Venus.

The role of the seemingly small animals that make up coral reefs in planetary processes emerges due to the cumulative impact coral have had, and currently do have, in influencing global climate by sequestering carbon. Emerging research regarding the evolution of the atmosphere on Earth versus Venus attributes an exceptional contribution from marine organisms. On Earth such organisms, including the evolutionary antecedents of coral, played a key role in sequestering carbon from the atmosphere, first by incorporating carbon as calcium carbonate to form their shells and, second, by depositing those shells on the sea floor upon their death, which accumulated to form the mass of limestone deposits across the globe. The Venus Syndrome hypothesis also holds that a central difference between Earth and Venus was that Earth's marine organisms removed enough carbon dioxide from the atmosphere to keep the planet cool enough to avoid runaway climate change. As Robin McKie, science and technology editor for the *Observer*, argues:

> Thus conditions on early Earth, which were very slightly cooler than Venus, prevented our planet from losing its water. This in turn played a critical role in permitting the formation of life forms that then reduced levels of greenhouse gases in our atmosphere and kept our planet habitable. (McKie 2015)

The depictions in *Coral: Rekindling Venus*, most of which are from the Great Barrier Reef, take on a deeper significance in light of their historical, contemporary, and future planetary agency. This reef, the only living organism visible from outer space, exhibits a sensitivity to interplanetary influences that the microscopic scale of its polyps and zooxanthellae belies. During the reef's annual mass spawning event, hundreds of coral species synchronize their spawning with the light of a full moon, water temperature and salinity, day length, and tide height. Amid these ecosystems vying year in and year out between the vicissitudes of local temperature versus the strength of moon light, are found also the accidental detritus from Cook's era of scientific and colonial exploration as well as materials deposited by deliberate interventions aimed at creating artificial reefs to replace reefs eroded by cumulative human impact.

Facing the Future

The exchange between Attenborough and Obama alludes to the role of coral in precipitating such deliberate planetary interventions. As the dominant "canary in the coal mine" of ocean warming and ocean acidification, coral inadvertently drives arguments for geoengineering as the means by which biota dependent on Holocene-type conditions may continue to flourish in the Anthropocene. Highly reputable global scientific research organizations are now researching geoengineering as a mechanism to mitigate warming and acidification. Scientists who advocate such research include Paul Crutzen, who brought the term "Anthropocene" into the public domain, and Sir Martin Rees, cosmologist and astrophysicist and former president of the Royal Society, the same organization over which Joseph Banks presided at the advent of the Anthropocene.

Proposals for planetary-scale geoengineering, from ocean fertilization with iron to stratospheric sulfur particle injection, are mirrored by local-scale proposals, such as cloud brightening, use of buoyant shade cloth to form sunshades over reefs, and use of electrical current to stimulate coral growth. These large-scale proposals are analogous to the invisible genetic engineering and visible PVC substrate of the artificial coral reef sample in the Cabinet. The larger proposals are seemingly invisible as they intentionally modify the chemistry of the atmosphere to make what is effectively a planet-scale sunshade of sulfur particles, much like modifying the genetics of coral through synthetic biology. The more local proposals are highly visible, as they form sunshades of clouds or cloth over regional reef sections, much like providing life-support structures of the artificial coral reef through PVC pipes.

A sample of artificial coral reef in a Cabinet of Curiosities for the Anthropocene, encountered by a viewer dead ahead and at eye level, is an emotionally layered Anthropocene object for an entwined present and future of human-nature hybridity. To gaze at the animal of soft coral tissue, the plant of the algae, the hard structure of minerals secreted by the corals, all together in suspended animation over a substrate of PVC, immersed in formaldehyde, and encased in glass, is to collectively see the Anthropocene writ large and small from the scale of the planet to the scale of a cell. It is to reconstruct the past and to forecast the future, and in so doing, to grapple with the profound ethical conundrums that have brought conservation biology and environmental engineering into alignment wherein the existence of multitudes of species may depend on what intentional experiments humans exert through synthetic biology and geoengineering. This artifact in its cabinet frames one of the great debates that will dominate for decades to come: after such inadvertent engineering of the environment, will intentional synthetic biology and geoengineering be the means by which life on Earth will continue onward toward the next transit of Venus in 2117?

NOTE

1   The meeting between Attenborough and Obama was filmed and presented by the British Broadcasting Corporation as *David Attenborough Meets President Obama*, directed by Ruth

Roberts. Available at https://www.youtube.com/watch?v=NZtJ2ZGyvBI. Subsequent
quotes are from the same source.

BIBLIOGRAPHY

Delaney, J. 2010. *Strait Through: Magellan to Cook and the Pacific, an Illustrated History.*
Princeton, NJ: Diane Publishing.
McKie, R. 2015. "Scientists Hope Venus Will Give Up the Secret of How Life Evolved on
Earth." *The Observer*, April 14.
Royal Society. 2009a. *Geoengineering the Climate: Science, Governance and Uncertainty*, September 1.
———. 2009b. *Symposium on Opportunities and Challenges in the Emerging Field of Synthetic Biology: Synthesis Report*, July 31.
Slezak, M. 2016. "Sections of Great Barrier Reef Suffering from 'Complete Ecosystem Collapse.'" *The Guardian*, July 21.
van Oppen, M. J. H., J. K. Oliver, H. M. Putnam, and R. D. Gates. 2015. "Building Coral Reef
Resilience through Assisted Evolution." *Proceedings of the National Academy of Sciences* 112
(8): 2307–13.

Cryogenic Freezer Box

Elizabeth Hennessy

A key strategy of environmentalism in the Anthropocene is to freeze life. In the midst of a sixth mass-extinction event—death on a scale that has not occurred since the disappearance of dinosaurs 65 million years ago—cryogenic zoos, or so-called frozen arks, have become an important technology for protecting the earth's biodiversity. In safe deposit boxes like the one shown in the Cabinet of Curiosities (plate 8), cryogenic freezers cooled by liquid nitrogen store fragments of charismatic fauna: DNA in the form of tissue samples, cell cultures, semen, ova, embryos, and blood. Such boxes are the interior scaffolding of modern biological archives.

But these frozen archives are more than just places to safeguard life in a threatening time. As a response to the Anthropocene, the cryogenic freezer box holds the tantalizing promise of shattering the finitude of extinction. This is the goal of "de-extinction" of species, an emerging and much-hyped wildlife restoration strategy made possible—in theory—by this archival method by which fragments of life are stored neither dead nor alive, but frozen in between. As scientists explained in a recent National Geographic Society "TEDx DeExtinction" conference, they would use DNA from extinct species—passenger pigeons, great auks, Tasmanian tigers, even wooly mammoths—to bring them back to life. Environmentalist Stewart Brand, who helped organize the event, emphasized the emotional appeal of such a possibility. He admonished the audience not to mourn lost species, but to organize—as have he and his wife, former biotech CEO Ryan

Phelan. The pair leads Revive & Restore, an NGO that is putting tech and venture capitalist philanthropy behind the promise of reversing extinction.

Unlike the demise of the dinosaurs, this extinction even cannot be blamed on an asteroid. Environmentalists position human agency as having a dual role in the Anthropocene—both culprit of environmental destruction and potential savior of lost life. Cryogenic freezer boxes encapsulate both regret for biodiversity loss and faith in science and technology to deliver life from the shambles of massive environmental crisis. They store life not as a remembrance of the past, but as an insurance policy for the future. If people have caused these extinctions, the thinking goes, then we should find a way to fix them. As Brand has put it: "We are as gods and HAVE to get good at it" (2010, 20).

But will gods and arks weather the flood of the Anthropocene? If archives are a reflection of how societies organize their relationships to the world around them, then what can cryo-deposit boxes tell us about the social-natural relationships of this "human epoch"? Who gets to "play god"? Faced with climate change, rising oceans, and other Anthropocene crises, how do these "gods" choose who, or what, should be saved? And if scientists in elite laboratories were able to revive extinct species, where in the world would these animals belong once they left the safe haven of the archive? Is there space in the world today for revived animals even as conservationists extract threatened species from their environments to ensure their survival? I argue that cryogenic freezer boxes not only contain bits of the charismatic species conservationists deem worthy of protection but also hold the broader tensions at the heart of environmentalism in the Anthropocene.

The Parable of the Passenger Pigeon: From Mourning to Techno-optimism?

No species has received more attention as a candidate for de-extinction than the North American passenger pigeon. In the mid-nineteenth century, massive flocks composed of millions of passenger pigeons stormed across the skies above the Midwestern United States. The birds, perched three deep on sagging tree branches, were easy hunting and a popular, inexpensive food source at a time when nature was seen primarily as a resource for human consumption. But this resource did not last. By 1900 flocks of pigeons had dramatically dwindled, prompting speculation: had they taken off for overseas or had their numbers fallen because of human hunting? The death of the last passenger pigeon, Martha, at the Cincinnati Zoo on September 4, 1914, spurred a wave of mourning and cautioning against destructive overuse of nature (Price 1999). In 2014 commemorations of the centenary of the birds' extinction made the passenger pigeon a poster species for the Anthropocene, its story a parable for understanding changing relationships between humans and animals.

The demise of the passenger pigeon a century ago was one of the first events that drove home to the American public the idea that species could go extinct because of human actions. The birds' plight pushed environmentalists to reflect on human responsibility to the creatures with which we cohabit the earth. As Aldo Leopold wrote in 1946 to dedicate a statue commemorating the pigeon, "Perhaps this monument is not merely a symbol of the dead past, but also a portent of a different future. Perhaps we learn more from the dead than from the living" (quoted in Temple 2014, 12). A generation ago, Leopold inspired environmentalists to learn from past mistakes that jeopardized other creatures' existence.

Today, rather than mourn the species and attempt to learn from the dead, environmentalists like Brand are putting their faith in de-extinction to make amends for past human extirpation by bringing back living, flying, flapping flocks of passenger pigeons. The resurgence of the passenger pigeon in public discourse illustrates a shift in environmentalism. In 1946 Leopold rallied against the "cosmic arrogance" of the "power-science" of the day (primarily atomic energy and pesticide production), which he argued was taking human relationships with nature in a dangerous direction. But now, in the midst of declarations of the Anthropocene, faith in biotechnology to solve environmental crises stands in stark contrast to the technological skepticism that was central to mid-century environmentalism.

For Stewart Brand, an "engineer's bias" is necessary for getting good at acting as gods (Brand 2010, 21). De-extinction rests on the premise of modern biology that scientists can approach life through engineering principles—through an understanding of DNA as biological building blocks that can be broken into constituent parts, rearranged, and reconstructed according to human desires. One of Brand's collaborators, evolutionary biologist George Church of Harvard's Wyss Institute, explained that biology is becoming "an engineering discipline, with interchangeable parts, hierarchical design, interoperable systems, specification sheets" (178). This new biology holds the promise, as Church put it, of "taking evolution to places where it has never gone, and where it would probably never go if left to its own devices" (Church and Regis 2012, 12). Indeed, Church is developing what he calls an "evolution machine" that would allow him to redesign entire genomes.[1] Such an achievement could revolutionize the biotechnology industry with applications far beyond using fossil remains to resurrect extinct animals. What could be more god-like than engineering evolution?

Archiving Life

The hyperbolic promise of engineering life emerges from a change in the nature of biological archives. For centuries, natural history museums preserved dead specimens of the world's faunal biodiversity as taxidermied preparations and skins. Today, rather than kill threatened animals to add them to museum collections, conservation biologists take vials of blood and tissue samples from endangered species as well as from preserved museum specimens. They either store these scraps of life in laboratory freezers for short-term analysis or cryo-preserve them for long-term archiving in one of about a dozen cryo-zoos around the world. (Many more cryo-archives are dedicated to freezing seeds and plant life.)

The largest cryo-zoo is the Frozen Zoo® at the San Diego Zoo's Institute for Conservation Research. It is housed in a windowless room off a nondescript ground-floor hallway in an office building in the San Pasqual Valley. This converted office space is a modern Noah's ark. But rather than saving animals two by two, this ark holds more than 10,000 biological samples from more than 1,000 different species and subspecies. Biomaterials from these animals are frozen in stainless steel cylinders at –196 degrees Celsius—about four and a half times colder than an average winter night at the North Pole.

The impulse to collect has been integral to the production of Western knowledge of the natural world since the sixteenth century when Europeans brought home curiosities during an age of imperial exploration. Wealthy European nobles collected natural oddities to display in curiosity cabinets as a demonstration of their fortune, worldliness, and power over the natural world. Cabinets were designed to provoke wonder by juxtaposing incongruous objects and thus served as potent sites for the production of Western imaginations about the world. In the eighteenth and nineteenth centuries, the emphasis on aesthetics in curiosity cabinets morphed into concern for order and comprehensiveness in state-backed natural history museums. Public exhibits in metropole museums showcased taxidermied animals from remote corners of the world as a display of imperial strength, while separate scientific collections stored scores of preparations grouped by species as a basis for biological knowledge. What was not visible in either display were the often violent and bloody histories of colonial hunting through which museums acquired these animals—and which contributed to the threat of extinction (Whitaker 1996; Poliquin 2012; Rothfels 2008).

Collection remains integral to Western methods of saving and studying nature. Although the blood shed today is largely contained in vials, and international transfer of biological materials is governed by the Convention on International Trade in Endangered Species of Wild Fauna and Flora (CITES), an uneven geography of extraction and preservation remains a central feature of biological knowledge production. For modern study, animals are extracted not only from their ecological environments—often in the biodiversity-rich tropics—but also from their embodied forms. The tiny bits of material that go into the cryo-deposit box are meaningful mainly as data, as codes to be deciphered through modern technoscience. Frozen archives look very different than early modern curiosity cabinets and natural history museums. Modern biological archives order nature based not on aesthetics, but on a molecular understanding of life. The animals in the Frozen Zoo are unseen by—and would be unrecognizable to—public audiences.

This reorganization of the archive reworks the relationship between the science of life and life itself. Archiving life as data-dense building blocks means that cryo-preservation is not only about storing biological data, it is about remaking biology. The potential to revive past life overcomes a limitation of previous natural history preservation methods—what Rachel Poliquin calls "the impossibility of complete satisfaction" because taxidermied animals cannot come back to life to play out the scenes suggested in the dioramas that animated natural history museums in the twentieth century (2012, 8). Rather than the aesthetic pleasure derived through taxidermy as a mastery of stilled life, cryo-preservation allures with the Promethean promise of engineering life.

De-extinction is perhaps the most extreme use of cryo-preserved materials. It falls at the far end of a range of interventions that conservation biologists have called genetic rescue—the idea that they can increase the genetic diversity of species populations to make them more resilient to environmental change. The most common strategy involves improving the diversity of captive endangered species populations by transferring semen between animals in different zoos. De-extinction is a different beast entirely. Scientists working with Revive & Restore use historical DNA from preserved specimens of extinct species as the basis for re-engineering these animals. De-extinction would work—in theory, at this point—by editing DNA from extinct species into the genomes of closely related living species. But this remains the stuff of science fiction.

The instruction book for bringing a passenger pigeon back to life has yet to be written (see fig. 6). Although the title of a recent book, *How to Clone a Mammoth*, by evolutionary biologist Beth Shapiro, suggests otherwise, scientists do not yet

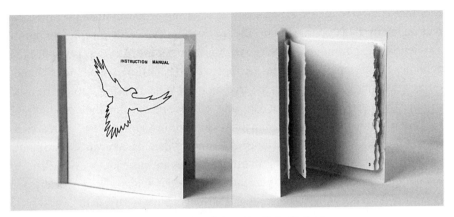

Figure 6. Instruction Book for Life. Courtesy of the artist, Helen Bullard.

have the technology to go from a genomic sequence to a revived animal. Along with Church, Shapiro's lab at the University of California at Santa Cruz is developing one of several potential methods by which to resurrect animals, but the processes remain in early experimental stages. For passenger pigeons, the plan is to splice pieces of the bird's genome into that of its closest living relative, the band-tailed pigeon. This, though, is a long way off.

So far, the most successful de-extinction project used a different method—"cloning" the Pyrenean ibex, or bucardo, by inserting preserved genetic material into the emptied ovum of a host mother goat. Of 154 embryos implanted into potential mothers, one ibex was born in 2003, but lived only seven minutes. She was malformed and unable to breathe properly. Ten years later, despite her death, the experiment is heralded as an early de-extinction trial, a proof of concept. But this biological feat came at the expense of sacrificed life. The ibex's story reflects what Thom van Dooren has called the "violent-care" of conservationist captive-breeding programs. Although intensive human effort has saved many endangered species, ethical examinations of breeding programs must not ignore the animal suffering and even death that often make these successes possible (van Dooren 2014). Even as de-extinction promises renewed life, it also promises sacrifices.

De-extinction experiments can be situated in a long history of tinkering with animals collected in archives. Curio collectors often enhanced natural wonders through artistry to dazzle audiences. Taxidermists have been known to piece together body parts from different animals to create mythical creatures like dragons, unicorns, and jackalopes. These chimeric creations blur the ancient metaphysical opposition between art and nature, the imaginary and the real. The resurrected ibex and dreams of revived passenger pigeons capture the wonder that was once central to the experience of encountering exotic animals in curio cabinets and natural history museums. But this wonder can easily morph into horror as scientists' experiments create monstrous forms of life. As scientists splice the genomes of different species in an attempt to bend back time, what kind of chimeric creatures might emerge from the cryo-archive? What thresholds might be used to judge the interplay of wonder, care, and violence that go into their production?

Perhaps, though, de-extinct animals are chimeric in another sense—unreal creatures of the imagination, the wild fancy of engineers' dreams. As even Shapiro points out, claims of exactly reproducing an extinct species are misleading: "We will never bring something back that is 100 percent identical—physiologically, genetically, and behaviorally identical—to a species that is no longer alive" (2015, 10).

A century ago, the last passenger pigeon, Martha, served as a parable of destruction. What then, today, is the tale of the cryo-deposit box? One way to think of this box is as a reverse of Pandora's, an archive that holds the Anthropocene's many mistakes as well as the prize of infinite life. At the bottom of this box, like Pandora's, is hope. But the question is whether this hope is ultimately dangerous, a hubristic promise that extinction can be fixed. Perhaps we cannot ultimately know—the future remains shrouded in the mist that surrounds the box. But this mist should not also obscure inquiry into the troubles of modern life that make the box necessary.

The task of the Anthropocene is not to fill a box with life and an instruction manual with technical directions for reversing extinction. Nor is it to abandon hope. Instead, the blank pages of the instruction manual can offer a different kind of guide, a space to reflect on a more complicated task: recognizing the human role in histories of environmental ruin, having the humility to know they cannot be fixed by extending the limits of life, and still daring to create a better future.

NOTE

1   See Jo Marchant, "Evolution Machine: Genetic Engineering on Fast Forward." *New Scientist*, June 22, 2011, https://www.newscientist.com/article/mg21028181-700-evolution-machine-genetic-engineering-on-fast-forward/ (accessed September 10, 2016).

BIBLIOGRAPHY

Brand, S. 2010. *Whole Earth Discipline*. New York: Penguin Books.
Church, G., and E. Regis. 2012. *Regenesis: How Synthetic Biology Will Reinvent Nature and Ourselves*. New York: Basic Books.

Poliquin, R. 2012. *The Breathless Zoo: Taxidermy and the Cultures of Longing*. University Park: Pennsylvania State University Press.

Price, J. 1999. "Missed Connections: The Passenger Pigeon Extinction." In *Flight Maps: Adventures with Nature in Modern America*, 1–55. New York: Basic Books.

Rothfels, N. 2008. *Savages and Beasts: The Birth of the Modern Zoo*. Baltimore: Johns Hopkins University Press.

Shapiro, B. 2015. *How to Clone a Mammoth: The Science of De-extinction*. Princeton, NJ: Princeton University Press.

Temple, S. 2014. "A Doubt We Have Gained." *The Leopold Outlook* 14 (1): 12.

van Dooren, T. 2014. "Breeding Cranes: The Violent-Care of Captive Life." In *Flight Ways*. New York: Columbia University Press.

Whitaker, K. 1996. "The Culture of Curiosity." In *Cultures of Natural History*, ed. N. Jardine, J. A. Secord, and E. C. Spray, 75–90. New York: Cambridge University Press.

Racism and the Anthropocene

Laura Pulido

116

Fossil fuels require sacrifice zones: they always have. And you can't have a system built on sacrificial places and sacrificial people unless intellectual theories that justify their sacrifice exist and persist: from manifest destiny to terra nullius to orientalism, from backward hillbillies to backward Indians.
Naomi Klein

To what extent has racism contributed to the Anthropocene? Although there have been heated debates on who, what, and where has caused the Anthropocene, there has been relative silence on the question of race. Discussions of liability are difficult under the best of circumstances, and including racism would certainly make them harder. But does that warrant ignoring it? Those parties most culpable for creating a new geologic era have actively sought to erase the power geometries that have produced it. Consequently, much of the Anthropocene discourse, especially emanating from the Global North, portrays it as a global problem that we have *all* contributed to. In response to such framings, Malm and Hornborg have suggested that the term "Anthropocene" is a misnomer, as it obscures the fact that only a relatively small percentage of the global population is actually responsible for and has benefited from the conditions that produced it. On those occasions when such disparities are raised, they typically emphasize the chasm between rich and poor, or "developed" and "developing" countries, as if the geography of wealth and power was somehow nonracial.

Coming from an entirely different perspective, leftists such as Jason Moore (2015) have suggested that the Anthropocene should be called the Capitolocene, attributing the problem to the prevailing economic system rather than individuals or countries. While almost all leftists acknowledge the unevenness of the Anthropocene, regularly citing colonialism, racism, and gender as important factors contributing to differential vulnerability, they usually treat racism as ancillary to capitalism.

While I would dispute that any single structure, event, or process created the Anthropocene, as one of the most profound social relations shaping the modern world, it is difficult to believe that racism has not played a part. Abundant research indicates that not only do many environmental hazards follow along racial lines, but also many of the meta-processes that have contributed to the Anthropocene, such as industrialization, urbanization, and capitalism, are racialized. I argue that the Anthropocene must be seen as a racial process. Certainly it is not solely a racial process—that would be a gross overstatement—but it has played an important role in both producing it and in determining who lives and dies. I examine how racism is embedded in the Anthropocene by focusing on several key issues: the evidence of racially uneven vulnerability and death; the form of racism at work; our general inability to acknowledge it; and the importance of history in coming to terms with the racial dimensions of the Anthropocene. I conclude by arguing that the racially uneven geography of death from the Anthropocene should be understood as a contemporary form of primitive accumulation.

While the Anthropocene is a broad term that denotes diverse forms of human impact on the planet, I pay particular attention to global warming. As the centerpiece of the Anthropocene, it has received by far the most attention and thus is the logical place to analyze how racism is, or is not, understood in the Anthropocene.

The Evidence: The Geography of Racial Vulnerability

Some might refute the idea that global warming has anything to do with race. After all, climate change is affecting the entire planet. Moreover, the state powers seeking to respond to climate change, such as the Conference of Parties (COPS), include global representation. Yet, when we look at who will pay the greatest cost, in terms of their lives, livelihoods, and well-being, it is overwhelmingly, to borrow a recently revived term from Vijay Prashad (2007), the "darker nations." While some may believe these are random patterns or accidents of geography, climate justice activists understand that they result from deep historical processes. They recognize that the rich, industrialized countries, which are disproportionately white, will escape with vastly fewer deaths. This is not due to any kind of racial animus but, in addition to historical processes, it is the result of a particular form of widespread contemporary racism, indifference.

It is a truism that we will *all* be impacted by the Anthropocene, but we will experience it differently. As Andrew Ross (2011) notes, there are those who will merely be inconvenienced and there are those who will die . . . with numerous positions between these two extremes. Those experiencing inconvenience will confront higher prices for food, energy, and water, as well as the discomfort of extreme heat as we transport ourselves from one air-conditioned environment to another. Others will have to contend with the loss of their homes, as the land on which they live is swallowed by rising sea levels, forcing them to move elsewhere. Still, others will find that their resource base can no longer sustain them and will have to migrate in search of a different land base or, more likely, a job in an overcrowded city. And then there are those who will simply die. Since 1990 more than 20,000 people have died from heat in India, culminating in May 2015 when over 2,500 died in a single month. In addition to dying from heatstroke and heat exhaustion, people will die from their inability to tolerate migration, hunger, thirst, and disease.

This differential vulnerability in terms of the haves and the have-nots is acknowledged by many, but the role of racism is generally overlooked. However, even a cursory glance indicates that it is overwhelmingly places occupied primarily by nonwhite peoples that will pay the highest price for global warming: death. The evidence for the uneven and unfair distribution of death is overwhelming. Figure 7 presents two different maps that highlight the uneven geographies of death from global warming. The top map depicts carbon emissions by country. What stands out is the bloated nature of the United States and Western Europe, and the contracted size of Africa. The bottom map shows estimated deaths produced by four likely health consequences associated with global warming: malaria, malnutrition, diarrhea, and inland flood-related fatalities. In this map we see the opposite: the Global North is greatly shrunk, while both Africa and parts of Asia explode.

Even within the wealthy parts of the world, the spatial distribution of risk, vulnerability, and death follows along pre-existing lines of racial inequality. In the United States, for example, researchers have found that the urban poor, which are overwhelmingly nonwhite, will die at the highest rates because of a lack of air conditioning. In places like California, which are leading the way in terms of climate mitigation, researchers have found that the "Cap and Trade" program, which encourages industries and firms to reduce carbon emissions over time via an emis-

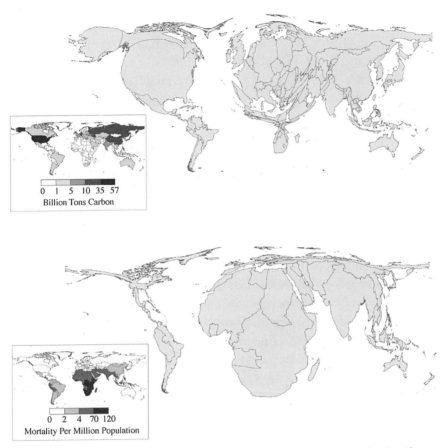

Figure 7. Comparison of global carbon emissions and mortality from global warming. Jonathan A. Patz, H. K. Gibbs, J. A. Foley, J. V. Rogers, and K. R. Smith, "Climate Change and Global Health: Quantifying a Growing Ethical Crisis," *EcoHealth* 4 (2007): 397–405.

sions market, has resulted in greater concentrations of air toxins for communities of color, thereby intensifying the environmental racism that already shapes California cities (Cushing et. al. 2016). Seen another way, Latinas/os and African Americans are subsidizing California's efforts to reduce global warming with their lives.

Disproportionate vulnerability can also be seen in indigenous communities, which are increasingly being called "frontline communities." Because they are land-based people, their livelihoods and way of life are extremely vulnerable to the Anthropocene, including species loss and change, flooding, and drought. Scholars such as Kyle Powys Whyte (2017) have argued that indigenous people are *already* living in dystopia, if one considers the ecological and social devastation of colonization, and that global warming is a continuation of a centuries-long apocalypse. Indeed, the ongoing nature of settler colonialism is evident in the fossil-fuel industry, which, assisted by states, continues to target these lands for extractive activities. Building on centuries of dispossession, these lands are often treated as "available" for taking by white settlers. Likewise, polluters and their fossil-fuel allies have long assumed that native peoples are weak and lack the political capacity to challenge them in a meaningful way. As a result, indigenous communities and their allies are engaged in intense battles across the Américas, from Canada to Peru. These communities not only wish to protect their lands, some of which are sacred, but also realize that continuing to extract and transport fossil fuels will place their communities and the entire planet in a more precarious position.

Evasion and Indifference

As previously noted, global leaders are well aware of the racial geography of the Anthropocene and have chosen not to act. Naomi Klein (2014), the only pundit who regularly discusses climate change as racial, recalls when the racist dimensions of our global strategy became painfully obvious. At the close of the United Nations climate summit in Copenhagen in December 2009, governments agreed to a global temperature increase of two degree Celsius. It was thought that two degrees would prevent global catastrophe. However, it was fully understood that two degrees would eliminate some island states and be absolutely disastrous for much of Africa. This is key: *knowingly* allowing large swaths of nonwhite, mostly poor people to die. Could we have decided to do otherwise? Yes. But as a global community we have declined to prevent this massive die-off. In response, a group of African delegates expressed their outrage, protesting: "We will not die quietly," "Two degrees is suicide," "1.5 to Survive," and "Death sentence for Africa." The delegates were not about to go quietly and wanted to ensure that everyone was aware of the import of their actions. This moment illuminated the racial geography of global warming, our lack of political will, our disregard for nonwhite and poor lives, and the deeply immoral nature of the Anthropocene. Since then, we have failed to even meet the goal of two degrees and are on target for four to six degrees Celsius temperature rise.

Given that our global leaders have condemned millions of people to death, we have to ask "why?" How are we able to make such a decision? Many would argue that the rich countries simply do not want to pay the additional costs of protecting and/or helping vulnerable countries. Certainly this is true. Few rich countries want their wealth siphoned off elsewhere, especially to nonwhite, poor places. This would result in fewer funds for domestic spending and the potential wrath of voters. It is uncertain how much political and moral capital it would generate in the global arena. As morally reprehensible as this may be, it is understandable. It's about money and power. We live in a capitalist world economy that places a premium on economic self-interest. But sentencing millions to die requires more explanation than simply economic self-interest. Such a powerful act requires an equally powerful ideology, as noted in the epigraph. And that is racism.

The global landscape of racism is vast and varied. The practice of racism and our understanding of it evolves over time and at any given moment multiple forms of racism are operating. The racism that undergird slavery and colonization is not the same as the racism that results in contemporary police shootings of Black people in the United States or the indifference we evince toward those who will die from the Anthropocene. During the conquest of the Américas, Europeans questioned whether Native Americans had souls, while Africans were considered to be a lower form of humanity. Over the centuries, through decolonization, the dismantling of slavery and apartheid, the development of human rights, civil rights movements, and other forms of antiracist struggle, racism has changed. Overt violence, legalized racial subordination, and racial animus have largely, although not entirely, been replaced by seemingly less intense, deliberate, and overt forms of racism.

Far more significant today is the *indifference* that characterizes the attitudes, practices, and policy positions of much of the Global North toward those destined to die. This indifference is a form of racism, because not only does it serve to reproduce racial inequality, but also this inequality enables the well-being of those destined to live. Ruth Wilson Gilmore defines racism as the "exploitation of group differentiated vulnerabilities to premature death in distinct yet densely interconnected political geographies" (2007, 28). Given this definition, while more than indifference was needed to create the conditions that produced such racially differentiated vulnerability, it is maintained by indifference. Like all other racisms, indifference is based on a devaluation of nonwhite lives and an overvaluation of white ones. We must recall that racism is first and foremost a relationship of power between two groups; it is not unidirectional. Thus, many in the Global North might assume the current valuation of white lives is the norm. But, as Lisa Cacho (2012) insists, racism is fundamentally a differential valuation. The devaluation of one group is predicated on the overvaluation of the other. While on the one hand, this may be obvious, on the other, many are hesitant to accept this basic truth. This general reticence can be seen in the slogan "Black Lives Matter" in which activists feel the need to proclaim the value of Black lives because they are so routinely devalued. Even the Gates Foundation, hardly a bastion of radical antiracist struggle, claims as part of its mission statement, "All lives have equal value." Again, they are responding to the general devaluation by the Global North of the lives of the Global South, where the foundation does much of its work. It is understandably difficult to accept the fact that we value lives differently—as this is contrary to contemporary human rights values, as well as the idea of racial progress. As a result, one of the primary responses to this contradiction is evasion, as we consistently seek to avoid addressing race.

Both evasion and indifference are visible in the objects and essays assembled in the Cabinet of Curiosities. While numerous essays acknowledge the uneven geography of the Anthropocene, there is little systematic analysis of racism. Several essays appearing in this volume readily acknowledge the historical significance of colonization, including Julianne Lutz Warren's "Huia Echoes," Trisha Carroll and Mandy Martin's *Davies Creek Road*, and Josh Wodak's "Artificial Coral Reef." Others gesture to the environmental injustice that Mexican farm workers experience with pesticides, but only one essay, Bethany Wiggin's analysis of the "Germantown Calico Quilt," seriously analyzes colonization and slavery. This is one out of fifteen essays. In contrast, a more common theme is to show how we *all* contribute to the Anthropocene, as seen, for example, in the wonderful story of concretes, or in essays in which the agent is never fully articulated. It is not my intent to critique the various authors, but rather to underscore the extent to which they reflect the larger dynamic of avoiding a serious engagement with racism. I appreciate the need and desire to highlight the fact that we are all active participants in the Anthropocene, but the lack of deep interrogation of one of the key inequalities shaping the geography of the Anthropocene must be pointed out. While some may argue that introducing racism into the discussion may create more problems than it solves—and it *would* create problems—it is essential that we account for *all* the processes that have contributed to the Anthropocene. Ideally, this volume would contain a healthy tension between the universalizing aspects of the Anthropocene, as seen by the concrete play, and by the specific dynamics responsible for the map presented in figure 7.

Besides our general desire to avoid acknowledging racism, there are genuine challenges to "seeing" race in the Anthropocene. One of the most important is the fact that over the last several decades a growing number of the "darker nations" have become major carbon emitters. The clearest example of this is China, which is currently the biggest emitter of carbon in the world. India is the third-largest emitter, largely due to its sheer population size. However, the emissions of both of these countries are relatively recent, exploding since 2000, and their per capita emissions are relatively low. For example, according to World Bank data, India's per capita carbon emissions is 1.7 metric tons, while Luxembourg's is 20.9. Such facts preclude drawing any clear racial lines in terms of emitters and victims. Yet, there are undeniable patterns in which the most vulnerable countries are overwhelmingly nonwhite.

Another challenge to seeing racism in the Anthropocene is the fact that in much of the world conceptions of racism have been constricted in order to minimize its perceived impact. Diverse strategies are employed to deny that racism is still a force in shaping the contemporary world, but what they all have in common is that they are predicated on decoupling race and larger material relations. This disconnection, Jodi Melamed (2011) argues, has facilitated restructuring conceptions of racism to fit particular political needs. It is the refusal to connect racially uneven outcomes with dominant attitudes, beliefs, practices, and structures that allows us to deny any possible connection between racism and the Anthropocene. Indeed, some ridicule the idea. For instance, the right-wing Breitbart News found it laughable that the *Guardian* sought to make a connection between racism and climate change (Williams 2016):

> The "reasoning" behind the outlandish hypothesis runs something like this. Begin with the unprovable premise that "Britain is the biggest contributor per capita to global temperature change." Next, assume that Britain "is also one of the least vulnerable to the effects of climate change," whatever that means. Finally, declare that "seven of the 10 countries most vulnerable to climate change are in sub-Saharan Africa." Climate change has just become a racial issue, wrought by selfish white people on unsuspecting blacks.

This quote invokes several mainstream strategies to delegitimize any effort to see the Anthropocene in racial terms. First, implicit is the assumption that racism is a conscious, hostile act. The author contracts the conception of racism so that other widely acknowledged forms of racism, such as white privilege or indifference, are irrelevant. Second, at work is a deeply ahistorical understanding of racism and, indeed, the world. Those who wish to avoid grappling with the legacies of previous racial formations, especially those based on more overt forms of white supremacy and violence, are deeply invested in ahistoricism. Countries like the United States, for instance, have developed numerous ideologies to defend the current social arrangement precisely in order to avoid acknowledging the racist past and its reverberations. Available ideologies include meritocracy, color blindness, multiculturalism, and postracialism, what Melamed has collectively called "state anti-racisms." Their power, she insists, lies in their ability to convince the general population that meaningful racial progress is being made, while simultaneously masking the violence of the contemporary racial order—which is precisely what is happening in the Anthropocene.

Historicizing Racism and Primitive Accumulation

While the Anthropocene is generally viewed as a potential catastrophe, one silver lining is that it forces us to reckon with history. This, in turn, provides an opportunity to reconsider the role of racism in shaping the present. Scientists have spent years deliberating when, if at all, the Anthropocene era should begin. In August 2016 the Anthropocene Working Group of the International Geological Congress announced its support for the formal recognition of the Anthropocene. While the Working Group proposed 1950 as one possible date when the era should begin, the question itself has launched a multitude of research initiatives to determine the appropriate marker, also known as the "golden spike." This is a deeply historical exercise. Regardless of what event or year is adopted as a marker, it is the deliberations themselves that are crucial as they offer an opportunity to revisit and reinterpret our collective past, and to hopefully come to a more honest accounting of how it is that we have created both the Anthropocene and its racial geography. As Gary Kroll notes in this volume, "the Anthropocene is less a geological epoch than it is a story."

Various dates and events have been suggested as a starting point, including the invention of the steam engine, human manipulation of fire, the Industrial Revolution, the plutonium fallout of the nuclear age, and the Great Acceleration of the 1950s. Each of these events offers a window into a whole series of racial dynamics that must be analyzed in order to ascertain the role of racism. In "Defining the Anthropocene," Lewis and Maslin (2015) offer two possible dates, 1610 and 1964. I would like to briefly explore 1610 because it allows us to confront the racist dimensions of the Anthropocene as seen through colonization, conquest, and primitive accumulation.

The authors chose 1610 because it marks European conquest and colonization of the Américas, one of the most singular events in human history. They cite two profound changes associated with 1610: transcontinental range expansion and a decline in carbon emissions. Transcontinental range expansion is evident in the spread of American species, namely corn, into Eurasia and Africa, as well as the transport of "Old World" species, such as bananas, to the Américas. This range expansion, while seemingly benign, both produced and was produced by a profound set of biologic, ecologic, and social changes.

The second major event associated with 1610 is a drop in atmospheric carbon dioxide. This was due to massive death in the Américas. Between 1492, when Columbus set sail, and 1650, it is estimated that approximately 50,000,000 people died in the Américas. This unprecedented die-off of humans resulted in a major decrease in farming, including a reduction in the use of fire for habitat modification. As a result, carbon was *not* released into the atmosphere through farming and other soil disruptions, ultimately resulting in a decrease in carbon emissions.

The "Columbian Encounter," as some euphemistically call it, is one of the most significant events in human history and led to vast economic, social, political, and ecological changes, including the previously mentioned dystopia for indigenous peoples. Lewis and Maslin's analysis, while certainly not intended to be political, illustrates that there is no escaping the political and power dynamics that have contributed to the Anthropocene. And though they never mention it, racism is a crucial feature of the events of 1610. One may wish to debate the merits of conquest and colonization, but there is little doubt that they relied on lethal force, state capacity, and a racial ideology of white supremacy. Despite whatever rosy stories we may tell ourselves, conquest was a bloody, violent affair. As Roxanne Dunbar-Ortiz reminds us, "people do not hand over their land, resources, children and futures without a fight" (2014, 8). The fact that most people died from disease rather than gunshot does not erase the racist dimensions of colonization. Indeed, colonists continued to wage war and genocide even after 90 percent of the population had been decimated. Nor can we lose sight of why Europeans were there in the first place—to conquer and claim the Américas, employing such self-serving legal justifications as papal bulls and the Doctrine of Discovery.

Contemporary Indigenous studies scholars, such as Jodi Byrd (2011), have pointed out that there are important distinctions between colonization and racism. For example, most antiracist activists desire inclusion, while decolonial activists desire autonomy and decolonization. Despite these distinctions, there is no escaping the fact that racism informs colonization. To claim another people's land as your own upon arrival; to kidnap people and force them into slavery or peonage and build an elaborate supporting apparatus; to eradicate another people's way of life; to steal the wealth and resources of another nation—these are breathtaking acts that require a powerful ideology to justify them. Such a sense of entitlement ultimately rests on a deep sense of superiority.

To understand how racism was harnessed for colonization, and subsequently capitalism, I draw on Cedric Robinson's concept of racial capitalism. Increasing numbers of critical ethnic studies scholars have begun challenging conventional Marxian analysis, which treats racism as incidental or, at the most, as an ideology to keep workers divided. Instead, Robinson (2000) argues that racism has been a constituent force of capitalism from the very beginning. Lisa Lowe, building on Robinson, explains, "the term racial capitalism captures the sense that actually existing capitalism exploits through culturally and socially constructed differences such as race, gender, region, and nationality, and is lived through those uneven formations; it refutes the idea of a 'pure' capitalism external to or extrinsic from, the racial formation of collectivities and populations" (2015, 150).

Seen in this way, racism informs contemporary capitalism and its antecedents, including primitive accumulation. For Marx, primitive accumulation was an early, violent stage of dispossession that was required in order to move into higher forms of human development. Marx observed, "The discovery of gold and silver in the Americas, the extirpation, enslavement and entombment in mines of the aboriginal population, the beginning of the conquest and looting of the East Indies, the turning of Africa into a warren for the commercial hunting of black-skins, signalized the rosy dawn of the era of capitalist production." Not only are scholars challenging traditional ideas regarding the relationship between racism and capitalism, but they are also challenging when capitalism is thought to have begun. While Marx himself saw primitive accumulation as distinct from capitalism, its predecessor, if you will, this is being forcefully challenged by the burgeoning literature on the history of capitalism. Is it accurate and meaningful to segregate capitalism from the relations that gave rise to it? Capitalism emerged, Marx continues, "dripping from head to foot, from every pore, with blood and dirt." Severing such violence and racism from capitalism is not only part of a larger ahistoricism, but it also serves to validate capitalism, as it is seen as less violent than its predecessor. However, primitive accumulation was essential to creating the initial surplus that subsequently allowed for the development of industrial capitalism. What is important for our purposes is that proto-capitalists, colonists, and Christians all drew on white supremacy as they went about the business of severing indigenous peoples from their land and labor.

It is important to recall that racism is not static. Europeans began with a particular conception of white supremacy but it was elaborated and enhanced as they went about their business of domination and exploitation. Though some dismiss primitive accumulation as irrelevant and/or ancient history, it is anything but. Decades ago Eric Williams documented how Caribbean slavery helped finance England's early industrial and financial development. Only now are greater numbers of scholars beginning to explore the economic and political implications of primitive accumulation. For example, Dunbar-Ortiz reminds us that in addition to the fact that the United States was built on stolen territory, as it took native land, it was placed in the public domain, sold, and generated funds to pay for an expanding military. This, in turn, supported overseas expansion, conquest, and empire. Recent books by Edward Baptist and Walter Johnson have also demonstrated how the profits of slavery contributed to contemporary US capitalism.

This is hardly the ancient past. These are the relations that birthed the modern world and which continue to shape it. As Marita Sturken reminds us, forgetting is a powerful form of memory. And we have put enormous energy into forgetting this history. It is only by re-engaging with it that we can appreciate the connection between the past and the present. Jack Forbes clarifies this relationship, noting that living persons are not responsible for what their ancestors did, but they are responsible for the society they live in, which is a product of the past.

While primitive accumulation helps explain the role of the past in producing the racial map of the Anthropocene, it is relevant for another reason. While many relegate primitive accumulation to the annals of history, the truth is that it is back with a vengeance. While one could argue that primitive accumulation never ended in the Global South, it has reappeared in the Global North. David Harvey argues that primitive accumulation has become a dominant form of accumulation in the contemporary period because the rates of profit have fallen so markedly. In response, capitalists have innovated and found new ways of producing profits and power across the world. Primitive accumulation, some argue, can be seen in the massive displacement from gentrification in many parts of the world, the 2008 housing collapse, and in the poisoning of Flint, Michigan's water. These contemporary forms of accumulation are violent forms of taking, as people lose their lands, lives, and livelihoods. Both old and new forms of primitive accumulation require enabling ideologies. And though there have been important changes, racism, especially indifference, remains an important one. Allowing millions to die ensures the wealth, prosperity, and convenience of rich countries, as well as powerful industries and firms. By not intervening in the processes that will produce massive death, they avoid burdensome regulations; they bypass a disruptive, rapid shift away from fossil fuels. Instead, they carry on as usual, working to maintain their profit levels despite the fact that the physical environment is shifting beneath their feet.

In the face of such dire circumstances, how has the global community responded? We craft global accords, such as the Paris Climate Agreement, which not only rely on voluntary reductions, since countries refused to adhere to the mandatory reductions of the Kyoto Protocol, but largely ignore the historical contributions of cumulative carbon emissions. A negotiator for the Seychelles reported, "The idea of even discussing loss and damage now or in the future was off limits. The Americans told us it would kill the COP." While the global community congratulates itself on achieving what is politically possible, we cannot overlook the anemic nature of the agreement considering the magnitude of the problem. It will not avoid the death of millions—because they simply do not matter.

BIBLIOGRAPHY

Baptist, E. 2014. *The Half Has Never Been Told*. New York: Basic Books.

Byrd, J. 2011. *Transit of Empire: Indigenous Critiques of Colonization*. Minneapolis: University of Minnesota Press.

Cacho, L. 2012. *Social Death: Racialized Rightlessness and the Criminalization of the Unprotected*. New York: New York University.

Cushing, L. M. Wander, R. Morello-Frosch, M. Pastor, A. Zhu, and J. Sadd. 2016. "A Preliminary Environmental Equity Assessment of California's Cap-and-Trade Program." Program for Environmental and Regional Equity, University of Southern California.

Dunbar-Ortiz, R. 2014. *An Indigenous Peoples' History of the United States*. New York: Basic Books.

Gilmore, R. W. 2007. *Golden Gulag*. Berkeley: University of California Press.

Glassman, J. 2006. "Primitive Accumulation, Accumulation by Dispossession, Accumulation by 'Extra-Economic' Means." *Progress in Human Geography* 30 (5): 608–25.

Johnson, W. 2013. *River of Dark Dreams*. Cambridge, MA: Belknap Press.

Klein, N. 2014. "Why #BlackLivesMatter Should Transform the Climate Debate: What Governments Would Do If Black and Brown Lives Counted as Much as White Lives." *Nation*, December 12.

Lewis, S., and M. Maslin. 2015. "Defining the Anthropocene" *Nature* 519:171–80.

Lowe, L. 2015. *The Intimacies of Four Continents*. Durham, NC: Duke University Press.

Malm, A., and A. Hornborg. 2014. "The Geology of Mankind? A Critique of the Anthropocene Narrative." *Anthropocene Review* 1 (1): 62–69.

Melamed, J. 2011. *Represent and Destroy: Rationalizing Violence in the New Racial Capitalism*. Minneapolis: University of Minnesota Press.

Moore, J. 2015. *Capitalism in the Web of Life*. New York: Verso.

Prashad, V. 2007. *The Darker Nations: A People's History of the Third World*. New York: New Press.

Robinson, C. 2000. *Black Marxism: The Making of the Black Radical Tradition*. Chapel Hill: University of North Carolina Press (originally published 1983 by Zed Press).

Ross, A. 2011. *Bird on Fire: Lessons from the World's Least Sustainable City*. New York: Oxford University Press.

Sturken, M. 1997. *Tangled Memories*. Berkeley: University of California Press.

Whyte, K. P. 2017. "Our Ancestors' Dystopia Now: Indigenous Conservation and the Anthropocene." In *Routledge Companion to the Environmental Humanities*, ed. Ursula Heise, Jon Christensen, and Michelle Niemann. London: Routledge.

Williams, E. 1994. *Capitalism and Slavery*. Chapel Hill: University of North Carolina Press.

Williams, T. 2016. "Guardian: Global Warming Is 'Racist'" Breitbart, September 8. http://www.breitbart.com/london/2016/09/08/guardian-global-warming-racist/ (accessed January 23, 2017).

Sabotaging the Anthropocene

Or, In Praise of Mutiny

Marco Armiero

> Since most of history's giant trees have already been cut down, a new Ark will
> have to be constructed out of the materials that a desperate humanity finds
> at hand in insurgent communities, pirate technologies, bootlegged media,
> rebel science and forgotten utopias.
>   Mike Davis, "Who Will Build the Ark?"

Let me be clear. Hopefully, there will be no room for misunderstandings. I am not at ease with the Anthropocene. I much prefer to ally myself with those who speak of the Capitalocene and remain skeptical of the universal "we," which is so central to Anthropocene narratives. The shift from the Anthropocene to the Capitalocene calls into question both the subject and the object of the story; it challenges the idea of a universal humanity, whereby all human beings are equally responsible for and affected by global environmental change, and the genealogies or, we might say, the causal historical connections that have led to the present crisis. Many scholars have written on the topic. A lively debate now divides the field between Anthropocene and Capitalocene supporters. Actually, the "ocenes" have been multiplying in keeping with the Great Acceleration, thereby uncovering numerous characteristics and causes of the current crisis—and maybe also the vanity of scholars or simply their desperate need to be quoted in order to survive in the Publish-or-Perish-ocene of the current university.[1]

But the debate over the capitalistic origins of the Anthropocene goes far beyond the walls of the academy; in meetings such as the People's Climate Change Conference—the name contains a program—activists openly blamed capitalism for climate change. In a document produced in 2010, the working group on "structural causes" wrote:

> Capitalism as a patriarchal system of endless growth is incompatible with life on this finite planet. For the planet, every alternative for life must necessarily be anticapitalist.[2]

The eco-socialist coalition "system change not climate change" expresses the same concept when it states that "the current ecological crisis results from the capitalist system, which values profits for a global ruling elite over people and the planet."[3] Naomi Klein (2014) speaks of a global blockadia, connecting thereby the variety of local, place-based struggles against invasive and polluting infrastructures that are emerging everywhere from Canada to Greece. Stopping fracking, gold mining, or oil extraction is a reality check that proves there is no obvious "we" in the Anthropocene. This does not mean that there is no "we" on the blockades of the Capitalocene. But it is not a predetermined "we"—be it the universal hubristic species of the Anthropocene or the local ecological community of some simplistic counternarratives. It is, rather, a "we" that is always in the making, a "we" made of alliances and mutual recognition. As Hardt and Negri have written in their attempt to delineate what they call the multitude, "Class is the product of class struggles. Class is a constituent deployment, a project" (2001, 104). In this sense, the "we" does not precede the conflict but it emerges from it. The experience of oppression and resistance produces subjectivities and draws the permeable borders of belonging. One can look, for instance, at the Red Warrior Camp in North Dakota where activists from Black Lives Matter have joined forces with the indigenous people protesting against the Dakota Access Pipeline. As the Black Lives Matter's website states:

> America has committed and is committing genocide against Native American peoples and Black people. We are in an ongoing struggle for our lives and this struggle is shaped by the shared history between Indigenous peoples and Black people in America, connecting that stolen land and stolen labor from Black and brown people built this country.[4]

A different "we" is built at the Red Warrior Camp, as is the case everywhere where checkpoints seclude spaces of exploitation from those of profit. It is a "we" that goes beyond racial/ethnic identities while overcoming the Anthropocene's narrative of naturalized species belonging. This is precisely the difference between saying that Black Lives Matter or All Lives Matter. I share Judith Butler's point on this: achieving the universalism of "all lives matter," or in a broader sense the communal "we" of the Anthropocene, is not a natural fact but the product of militant struggles.[5]

The universal "we," I argue, is the most powerful narrative of the Anthropocene. And it trades on the compelling metaphor of the ship. "We are all on the same boat," says the refrain, delivering the easy one-worldism message that appeals sentimentally to the good will of each individual. The ship metaphor implies two simple things: the shared destiny of the people on board and the finiteness of their world. On a ship, and in the Anthropocene, there is no way out. The ship must survive the perils of the ocean. The passengers must make it work because they do not have any other options. At the moment, it is only in Hollywood, the planetary factory of disciplined imaginaries, where we have the option to leave the ship when it has failed and colonize a whole new frontier in space.

The ship is not a new metaphor to represent the special combination of humans and the environment on earth. The Ark, Spaceship Earth, and the lifeboat have often been employed to describe the human condition on the planet. The ship metaphor speaks of both salvation and ruin, exploration and haphazard wandering. Even today, the giant seed bank located in the extreme north on the Svalbard Islands, meant to insure humanity's future against the loss of genetic diversity, has been often called the Ark. The same name was given to a pan-European network dedicated to the "rewilding of Europe." What else could represent the effort to save life on Earth if not the mother of all ships, the biblical Ark, which once had preserved the planet from the deluge? But Noah did more than preserve nature. He built the ship that charted a new future for living beings on Earth. The Anthropocene is an age of creation rather than of rescue.

In the late 1960s the ship metaphor, as historian of science Sabine Höhler shows, was enlisted as a rhetorical weapon in the arsenal of environmentalist discourse (Höhler 2015). In truth, it was a rather different kind of ship, more in tune with the technological dreams of those times. Spaceship Earth was the image employed by the economist Kenneth Boulding in a speech delivered on March 8, 1966, at a Resources for the Future Forum on Environmental Quality in a Growing Economy. Boulding was not referring to boats and seas, but to spaceships launched into the vastness of the universe. He was not the only one to use that metaphor. In 1968 R. Buckminster Fuller published a book with the futuristic title *Operating Manual for Spaceship Earth* based on the same image. Nonetheless,

Boulding is often mentioned in relation to this metaphor and for his influence in the foundation of ecological economics. For Boulding, Spaceship Earth embodied the limitedness of the planet and the necessity to think of it as a closed system in which people and resources must be in balance. Maybe even more than the (space)ship per se, Boulding's metaphor of astronaut vs. cowboy economies gave the sense of his vision of a limited planet. Unlike the cowboys, always looking for a Western frontier to exploit, the astronauts know that there is no other space where they can move to if all the resources on the ship are exhausted.[6] This sense of ecological limits was the main message of Boulding's ship parable and it acquired a growing relevance in the following years with the iconic image of the earth from space (1968), the Limits to Growth report (1972), and the oil crises (1972 and 1979).

In the early 1970s the American ecologist Garrett Hardin took issue with proponents of Spaceship Earth. "The spaceship metaphor," Hardin argued, "can be dangerous when used by misguided idealists to justify suicidal policies for sharing our resources through uncontrolled immigration and foreign aid. In their enthusiastic but unrealistic generosity, they confuse the ethics of a spaceship with those of a lifeboat" (1974, 38). The rich nations of the world, Hardin suggested, might be seen as lifeboats, surrounded by the poor swimming at sea. Who should the rich, first-class passengers let on board? None, argued Hardin. To do so would swamp the boat and everyone would die. "Complete justice, complete catastrophe," quipped Hardin.

Many have rightly criticized Hardin's argument. Scholars have pointed out the suspect absence of any historical explanation of the richness of the few and the poverty of those in the sea. Others have underlined Hardin's mistake in looking at those lifeboats as isolated systems, as if the wealth of those on the boats and the poverty of those struggling in the water were not intricately connected. Obviously, radical scholars have denounced his naturalization of injustice—someone happens to be wet, others dry and sound on the boat. Carrying capacity—the maximum population that any given environment can sustainably support—has often be used as a scientific base for anti-immigration sentiments. Hardin himself was directly involved in an anti-immigration think-tank, the Federation for American Immigration Reform, and a passionate advocate for limiting immigration in the United States. Although morally and politically disturbing, Hardin's lifeboat ethic did refer to the recurring trope of the ship as a metaphor of the human condition on a finite planet. In its inhumanity, Hardin's lifeboat ethic in fact most closely resembles the metaphor of life and death on this planetary ship where not all passengers experience travel or possible shipwreck in the same way.

Indeed, Hardin's draconian proposal to let people die in order for the rich to live harkens to a real tale of shipwreck in the modern age.

The *Titanic* story is the ultimate tale made of steel and celluloid about modernity and its limits. The story is well known: on April 10, 1912, progress set sail from the harbor of Southampton. It was a black, gigantic ship, equipped with the best technologies of the day: a high-power radiotelegraph transmitter and remotely activated watertight doors. It also had all the amenities that made it a demonstration of the technological and economic power of Anglo-Saxon society. Nonetheless, as sometimes happens with high-technology tools and luxurious gadgets, the *Titanic* lacked a more basic device: enough lifeboats to carry the entire crew and passengers on board. So, when on the night of April 15 the ship met an iceberg, almost 1,700 people died. Among all the ship metaphors, the *Titanic* at first glance best seems to fit the human condition in modern times. With literary scholar Steven Mentz we can see the shipwreck not as an apocalyptic projection of what might come but as the very essence of humans' experience in the present time. Indeed, tales of the Anthropocene are made concrete in the *Titanic*'s story. The crew and passengers boarded "an unsinkable ship," a myth fabricated through blind faith in technology and an abiding trust in experts. Even the dynamic of the disaster evokes some common assumptions about how the coming apocalypse will materialize: the inability of the leadership to foresee the cataclysm and take safety measures in time; the passengers, distracted and incredulous, dancing in the ballroom while the ship encountered its fate. If one adds the fact that in the end it was an iceberg that struck back against the hubris of technological perfection and unsinkable modernity, the resonance with the Anthropocene condition becomes striking.

But the *Titanic* is also a perfect Anthropocene tale because its circulation in cultural memory occludes at least as much as it reveals. About 75 percent of the first-class passengers survived, while only 20 percent of the passengers in steerage made it out alive. Although the racial policies of the White Star Line regarding the passengers are rather unclear, evidently the company preferred to keep the "crew rosters as lily-white as possible."[7] Race factors into the *Titanic* story in interesting ways. The historian Steven Biel has shown how little interest the African American community had in an event that galvanized the attention of largely white, middle- and upper-class society across both sides of the Atlantic. In Biel's words, the *Titanic* reminds us that "what seemed universal to some was actually a matter of perspective" (2012, 108).

"Shine and the *Titanic*" became a popular toast among African Americans after the *Titanic* went down. The toast tells the tale of Shine, an imaginary black stoker on board the *Titanic*, who continually warns the captain of the impending disaster, but whose pleas are repeatedly brushed aside. Shine dives into the icy waters, outfoxes a whale, and safely swims to shore while the ship goes down. Shine's story resembles a mutiny or, perhaps, a carnival subversion of the usual power relationships; here an African American, who might have been denied passage, is clearly superior to the imbecile, white captain blind to the destiny of his ship. But mainstream memory built another narrative based on the sacrifice of white male officers who fired shots at the uncivilized crowd emerging from the hold of the ship as they tried to escape the rising waters below. As the ship went down, order and discipline needed, after all, to be maintained.[8] Indeed, class matters in the Anthropocene. Race matters in the Anthropocene. On the *Titanic*, there was no "we" before or after it sank.

But other maritime stories, which might appear at first blush less germane to the Anthropocene than the *Titanic*, come to mind. These are stories of mutinies. There are so many of them, some quite well known, as the *Bounty* or the *Amistad*—and Hollywood did help in elevating them to the *Titanic*'s fame—but others are almost absent in the collective memory, although they marked important turning points. Although I have always found the *Bounty*'s story extremely appealing to the collective imaginary of the mutiny, nonetheless, its colonial implications and questionable outcomes require us to look elsewhere. The Kiel mutiny in the autumn of 1918 has neither the glamor nor the notoriety of the *Bounty*; nonetheless, it is a fitting story for the Anthropocene. True, the Red Sailors of the German Navy did not conform to the global imagination of romantic rebels. Never mind. We need for our times mundane mutineers, saboteurs completely imbricated in the capitalistic reproduction of war and exploitation rather than heroes hiding on some remote islands.

Unlike the *Titanic*, which seems both dramatically isolated and drenched in nostalgia, the Kiel mutiny demonstrates the connections of the technoscientific complex of the sea and the social relationships producing it. The mutiny involved a large part of the German sailors serving on the High Seas Fleet of the Reich, a fleet clearly embedded in a network of imperial and class relationships and at the heart of the nation's military-scientific complex. The High Seas Fleet was supposed to embody the imperial ambitions of Germany, competing with the British Navy for the control of the sea and, in particular, of colonial shipping routes. While the *Titanic* tale is confined to the weaknesses of the ship, the mistakes of captains, the disinterest of the people on board, the Kiel mutiny reminds us that the ship must be understood within a web of social relationships that goes beyond the vessel. A warship, with its heavy cargo of discipline, expertise, and chain of command stands, like the *Titanic*, as a symbol of modernity. And on the face of this stands the mutiny. In October 1918, facing the suicidal order to attack the British Navy, the sailors of Kiel opted to mutiny, took over their ships and plotted a new exit trajectory from the imperial relations of war and exploitation in which they were enlisted to defend. As the High Seas Fleet was trapped by the blockade of the British Navy, so those sailors were trapped not just on their ships but also in the disciplinary order that gave sense to their condition. The trap had to be broken, but instead of striking out toward the enemies, the "red sailors"—as they were called—looked for alliances among subaltern groups in Germany aiming not only to control their ships, but to change radically the system that had produced both the High Seas Fleet and the war. Mutiny opens up a radically different future; it subverts the logic governing the ship, instead of trying to improve it. According to several historians, the Kiel mutiny ignited the revolution that brought the end of both war and monarchy in Germany. Others have questioned the political motivations of the Kiel mutiny, arguing that, after all, the sailors wished only to have better food and avoid a suicidal mission, as if improving the living conditions or refusing to die for the emperor were not political matters. One might wonder what is political, or even revolutionary, in the Anthropocene.

The metaphor of the Kiel mutiny—actually of every mutiny—challenges the naturalization of social relationships in the Anthropocene and its postpolitical character. In a time of experts' anesthetic suggestions and apocalyptic prophecies, the mutiny recovers the space of subversion, opens up the possibility for radical changes, and reclaims the right to fight against the duty to follow orders—or maybe blindly implement technocratic, well-informed blueprints. The mutiny explains that the problem is not the ship per se, but the social relationships it keeps afloat. One cannot save it from disaster because the vessel is actually a function of the disaster. Mutinying means to subtract oneself from the totalitarian logic of the ship, subverting the hierarchies governing both the people on board and the mission of the vessel. On the Anthropocene boat, the best strategy to survive is to disobey or, we might say, sabotage the order that is driving to the final shipwreck.

Ships and mutinies evoke the two opposite poles of work and sabotage, of making and unmaking. Work and sabotage stay quite invisible on the ship as well as in the Anthropocene. What moved the *Titanic*, the intermingled bands of muscles, engines, and coals, remained hidden in the bowels of the ship. The sweat and violence of capital exploitation of labor is not at the forefront of the mainstream vision of the Anthropocene. In the Anthro-obscene, to borrow from Ernston and Swyngedouw, what stays invisible is not CO_2 emissions or the future apocalypse but the present violence of capitalist expropriation of the work of nature and people. Humans might be the gods but not the workers of the new age. The Anthropocene can make us dream of planetary geo-engineering of the climate but leaves us blind to the present geo-mining of coltan, made possible through exploitative labor practices that take a heavy toll on the bodies of children. Our global eye stays blind on Rana Plaza in Dhaka, Bangladesh, the deadliest garment-factory disaster in history, where the industrial engines that produce climate change feeds itself on the death of hundreds of workers. Indeed, the machines fueling the Industrial Revolution have never only devoured the countryside and occupied the atmosphere but also devoured and occupied the bodies of workers. Daegan Miller's object, the monkey wrench, speaks not only of victims but also of subjects. Proposing the monkey wrench, Miller claims back the agency of workers in the making of the Anthropocene, and, as always the case, those who can make can also dismantle, undo, sabotage.

The mutiny is the ultimate sabotage; it dismantles not only a machine but the logic governing the functioning of the Machine. The mutiny defuses the discourse of "working together for the common good," but it also questions individual escape, the minimalist attempt to exit the machine and find redemption alone, outside it.

In their book *Commonwealth* Hardt and Negri speak of exodus as a first form of class struggle in the age of biopower, as a way to subtract oneself from the capitalistic relationships (2001, 152), but they also remind us that in the age of Empire there is no outside. Thus, while capitalism internalizes the outside through the occupation/control of the body and the colonial project, both themes at the center of this section of our volume, the mutiny constructs an inside exodus, which sabotages the biopolitical expansionist project striving to free both spaces and people.

As Karl Marx elucidated long ago, work connects the human body and external nature. Since he envisioned it as a metabolic relationship, we might even say that the boundaries between the two become blurry. How would the Anthropocene look if, instead of searching for its traces in the geosphere, scientists would look for them in the organosphere, that is, in the internal ecologies of humans? Strata of toxics have sedimented into the human body, arriving, according to the most recent studies in epigenetics, to be inscribed in the genetic memory of humans. The pesticide pump, which Michelle Mart and Cameron Muir have included in our Cabinet of Curiosities, is the icon of this embodied Anthropocene. It nebulizes the dream of the total control over external nature, revealing it to never have been external at all. As in the Anthropocene, pesticides are ubiquitous, imperial in their occupation of space and bodies. Nonetheless, as in the

Anthropocene, pesticides benefit and affect different groups of people unequally. A housewife fighting against homely insects might have been the central figure in sales campaigns for pump sprayers, but in reality it has been an army of farm-workers who have worked more extensively with pesticides and whose bodies suffer the consequences.

While capital internalizes the body through work, it also internalizes the extreme otherness of the colonial space. Questioning the reification of the Anthro-pocene leads to a different chronology, as the one Bethany Wiggin has committed to in her essay. The Germantown Calico Quilt she describes frames the Anthro-pocene into the longer process of capitalism's imperial expansion as proposed by, among others, Moore and Lewis and Maslin. The very material of which the quilt is made embodies the blend of race/class/nature relationships producing and reproducing the empire. Expropriation of land and humans, as plantations and slaves, made the economy of cotton possible. The Penn's Treaty represented on the Germantown Calico Quilt carries with it the rhetoric of the benevolent white supremacy and the reality of the imperial power. But that calico speaks also of the entanglement between private lives and public (hi)stories, local belongings and global networks. In Mrs. Smith's quilt the Anthropocene wraps itself around into intimate lives, proving, I argue, not so much the god-like nature of humans but rather the pervasive nature of capital.

NOTES

1   Just as a sample here is a short list of ocenes: the econocene (Norgaard), the Chthulucene (Haraway), the Naufragocene (Mentz), the Wasteocene (Armiero and De Angelis), Homo-geocene (Suckling).

2   https://pwccc.wordpress.com/2010/04/30/final-conclusions-working-group-1-structuralcauses/#more-1775 (accessed on February 12, 2016).

3   System Change Not Climate Change website, http://systemchangenotclimatechange.org/about (access on February 12, 2016).

4   Black Lives Matter website, http://blacklivesmatter.com/solidarity-with-standing-rock/ (accessed on September 5, 2016).

5   George Yancy and Judith Butler, "What's Wrong with 'All Lives Matter'?" *New York Times* 12 (2015): 156.

6   In Boulding's own words: "The closed earth of the future requires economic principles which are somewhat different from those of the open earth of the past. For the sake of pic-turesqueness, I am tempted to call the open economy the 'cowboy economy'; the cowboy being symbolic of the illimitable plains and also associated with reckless, exploitative, romantic, and violent behavior, which is characteristic of open societies. The closed econ-omy of the future might similarly be called the 'spaceman' economy, in which the earth has

become a single spaceship, without unlimited reservoirs of anything, either for extraction or for pollution, and in which, therefore, man must find his place in a cyclical ecological system which is capable of continuous reproduction of material form even though it cannot escape having inputs of energy" (Boulding 1966). http://www.ima.kth.se/utb/mj2694/pdf/boulding.pdf (accessed on September 5, 2016).

7   Rice 2003, 32–33. My discussion of the African American memory of the *Titanic* is based on this book and on Biel 2012.

8   Rice refers the apocryphal story of a black stoker who tried to stab a wireless operator (32), while at the Senate Inquiry on the disaster the Fifth Officer Harold Lowe justified the use of his revolver saying, "I saw a lot of Italians, Latin people, all along the ship's rails—understand, it was open—and they were all glaring, more or less like wild beasts, ready to spring. That is why I yelled out to look out, and let go, bang, right along the ship's side." Excerpt from Lowe's testimony are available at https://www.encyclopedia-titanica .org/gunshots-on-titanic.html (accessed on September 5, 2016).

BIBLIOGRAPHY

Armiero, M., and M. De Angelis. 2017. "Anthropocene: Victims, Narrators, and Revolutionaries." *Southern Atlantic Quarterly* 116 (2): 345–61.

Biel, S. 2012. *Down with the Old Canoe: A Cultural History of the Titanic Disaster.* New York: Norton.

Boulding, K. 1966. "The Economics of the Coming Spaceship Earth." In *Environmental Quality in a Growing Economy*, ed. H. Jarrett, 3–14. Baltimore, MD: Resources for the Future/Johns Hopkins University Press.

Davis, M. 2010. "Who Will Build the Ark?" *New Left Review* 61:29–46.

Ernston, H., and E. Swyngedouw. 2015. "Framing the Meeting: Rupturing the Anthro-obscene! The Political Promises of Planetary and Uneven Urban Ecologies." Position Paper Version 2. Presented at Conference at Teater Reflex, September 16–19, 2015, organized by KTH Environmental Humanities Laboratory, Stockholm. http://www.anthro-obscene .situatedecologies.net/framing.html.

Haraway, D. 2015. "Anthropocene, Capitalocene, Plantationocene, Chthulucene: Making Kin." *Environmental Humanities* 6:159–65.

Hardin, G. 1974. "Lifeboat Ethics: the Case against Helping the Poor." *Psychology Today*, September, 38–43, 123–26.

Hardt, M., and A. Negri. 2001. *Empire.* Cambridge, MA: Harvard University Press.

———. 2004. *Multitude: War and Democracy in the Age of Empire.* New York: Penguin.

———. 2009. *Commonwealth.* Cambridge, MA: Harvard University Press.

Höhler, S. 2015. *Spaceship Earth in the Environmental Age, 1960–1990.* London: Routledge.

Klein, N. 2014. *This Changes Everything.* New York: Simon and Schuster.

Lewis, S., and M. A. Maslin. 2015. "Defining the Anthropocene." *Nature* 519 (7542): 171–80.

Mentz, S. 2015. *Shipwreck Modernity.* Minneapolis: University of Minnesota Press.

Moore, J. 2015. *Capitalism in the Web of Life.* New York: Verso.

———, ed. 2016. *Anthropocene or Capitalocene? Nature, History, and the Crisis of Capitalism.* Oakland, CA: PM Press.

Norgaard, R. B. 2013. "The Econocene and the Delta." *San Francisco Estuary and Watershed Science* 11 (3).

Rice, A. 2003. *Radical Narratives of the Black Atlantic.* London: Continuum Books.

Suckling, K. "Against the Anthropocene." Immanence: Ecoculture, Geophilosophy, Mediapolitics (blog). http://blog.uvm.edu/aivakhiv/2014/07/07/against-the-anthropocene/.

Laboring

On Possibility

Or, The Monkey Wrench

Daegan Miller

The monkey wrench. Almost nobody uses one anymore, except as an aid to nostalgia: perhaps you've seen one hung on the wall of a country diner, a memento of the unremembered simplicity of pie-and-coffee times; or maybe you've seen them displayed proudly—three for five dollars—as knickknacks at a local flea market. Or, if you're of the activist persuasion, you might recognize the wrench as it crosses a stone club on the insignia of the direct-action, antimodern environmental group Earth First! But you almost certainly won't see a monkey wrench at work—they were replaced, in the waning decades of the twentieth century, by lighter, more precise wrenches, and today, if it is used at all, the monkey wrench is mostly a tool for assembling supposed histories made of air (plate 9).

Perhaps this was all genetically predetermined: the tool has no birthdate, no clear nationality, no uncontested paternity. Though some credit Charles Moncky, a British emigrant to the United States, with conceiving the thing sometime in the mid-nineteenth century, others hold that the American machinist and factory owner Loring Coes delivered forth the tool—or appropriated the design from an employee named Monk . . . or bought the design from a man named Monckey. It *is* true that Coes held an 1841 patent for what he called the "screw wrench," and it's also true that Coes's wrenches became the industry standard for what an adjustable wrench ought to be, though there's a debate among American tool enthusiasts, flushed with nationalist anxiety, whether Coes's patent makes the monkey wrench American, or whether the thing is simply an updated version of the eighteenth-century English carriage wrench. The debate may be moot, because it also seems likely that the term "monkey wrench" predated Coes—collectors have turned up references to the tool as early as 1807, a date of which the Oxford English Dictionary is skeptical—and it seems that by the time the *Natchez Daily Courier* ran an advertisement in 1838 for a local hardware store, the tool was well known enough to need no explanation. Maybe the wrench is named for an inventor. Maybe not. Maybe it's a bastardization of "moving wrench," as some believe, or a tongue-slipped consonant away from its technical classification as a "non-key wrench." Non-key; non-key; non-key: say it fast enough, frequently enough, and it's possible that evolution occurred. Or maybe there's an act of recursive instrumentalization at work: since the seventeenth century, manual laborers have been sneered at as monkeys—powder monkeys loaded cannons and grease monkeys maintain cars. A plumber with whom I once worked on a Rockefeller estate (I installed lawn sprinklers during breaks from college to help pay for school) referred to all of us with rough hands, bitterly, as "dumb wrenches." Maybe monkey wrenches are the only tools simple enough for the least of us to wield.

Whatever its history might be, what a monkey wrench is is less important than what it does. I have one, now, on my desk, and I've come to think that historical mystification is the work of the stubborn tool itself. Once used everywhere lithe human muscle struggled against iron intransigence, the monkey wrench had a hand in building the entire towering, now tottering mechanical skeleton of the industrialized, modern world—of the Anthropocene. Perhaps the wrench's latest act is to refashion history by twisting the historian's linear, rational, absolute time into its own likeness: like its past, like our future, the monkey wrench is literally a question mark.

And so the wrench asks us: what is this Anthropocene—this age in which, the term's inventors tell us, man makes everything anew; this age whose occluded dawn is pegged to the very years of the humble wrench's unrecorded birth? An imprecise tool—its toothless jaws are only grossly adjustable—the monkey wrench nevertheless firmly catches the slippages of others: the casual sexism of defining an era as "man's" and the injustice of assuming that humankind, Exxon-Mobil board member and migrant laborer alike, is equally responsible for the industrial revolution; for the proliferation of wealth's byproduct, carbon dioxide; for the great die-off of flora and fauna marking this, the sixth age of extinction; for the poisoning of our atmosphere, our water, our soil, our bodies. The monkey wrench catches the slippage of how the name Anthropocene calls attention to what humans have done to the world while ignoring what we've done to each other, and holds it still for a moment, still enough for us to puzzle over the oddity of our situation, of an accidental age named for human ineptitude.

And so the wrench allows us, if we pause in our work for a moment longer, to consider inequality—*whose* labor built the Anthropocene? *Whose* labor laid the rails, fitted the pipes, shoveled the coal, felled the trees, grew the grain, picked the cotton, slaughtered the cattle, sailed the ships, forged the iron, drilled the wells, trucked the oil, poured the concrete, assembled the engines, mined the ore, strung the wires giving light, motion, form, and strength to the Age of Man? *Whose* labor brought many millions of tool-handling workers into the world? *Where* did all this work happen? What parts of the world were looted for their wealth—their precious ores, soils, trees, and animals—and what parts of the world have become dumping grounds for the toxic effluvia of industry? Which parts of the world will be saved from the worst effects of the Anthropocene, and which derelict Atlantises will be left to slip beneath rising, acidified seas?

The monkey wrench reminds us that on the other side of every cost stands a profiteer, and when once again held in a warm human hand, the wrench confronts us: *who* profited from its work and *who* has paid the costs? I bought my first monkey wrench on eBay for $15.00 (it now lives in the Deutsches Museum), and it was once owned by someone who stamped his initials—MAM—into it in two places. It was a valued instrument. No doubt the wrench also left its impression on MAM: monkey wrenches were notorious for slipping under high pressure, just when their users most needed their jaws to bite securely on a nut. When it slipped, workers got hurt—bloody knuckles and purple bruises, and I wonder: what was MAM paid, and was it compensation enough for his spilled blood? How much of the sweat from the monkey's brow went to sate the enormous appetite of another with cleaner, softer, probably whiter hands?

If, as Paul Crutzen and Eugene Stoermer argue, the Anthropocene was born amid the atmosphere-altering exhaust coughing from the coal-fired steam engines powering the Industrial Revolution, engines that within a hundred years would start firing on refined petroleum, then it's also true that the Anthropocene's other parent was the exhausted worker toiling away at those same machines that devoured the countryside and, all too often, the humans who tended them. There was a third, of course: he who converted humans and nature into resources, merely assets awaiting conversion into capital.

This is, after all, how profit works in the Age of Man.

Words, like wrenches, are tools that help us get a grip on the world, and names are micro-narratives, stories that ascribe responsibility, advocate for morality, and seed possibility. So perhaps Anthropocene is the wrong term (or the right one for deflecting too-pointed questions). Perhaps not all humans bear equal guilt, just as not all have reaped equal rewards. Perhaps Plutocracene, the Age of the Wealthy, is a better fit, or Kleptocracene, the Age of Thieves, the age when the productivity of the earth and all its living things was stolen away.

Of course, tools are only infrequently aids to past reflection. Most often, we use them to build, and they always anticipate future action. If the Anthropocene is the ironic result of a scientific, technological, economic, and political drive to control nature and humans alike, an age that was supposed to usher in great prosperity, but which paradoxically impoverishes everything, then the wrench asks us what we will do about it. Yet no tool, even one as uncomplicated as the monkey wrench, is simple.

If it seems built to turn a bolt, its hammerhead testifies that monkey wrenches were also used to bash a stuck bolt. Tools can demolish. By 1907 the monkey wrench found itself a comrade in industrial sabotage, followed shortly by a grammatical shift: the noun became an unspaced verb—to monkeywrench—with an activist connotation: if you don't like your master's world, tear it down. In 1975 the monkey wrench emerged as a potent symbol for environmental, antimodern direct-action when the American anarchist and environmentalist Edward Abbey set loose his novel *The Monkey Wrench Gang*. The book—dedicated to the British loom-smashing critic of the Industrial Revolution, Ned Ludd—revolves around the dream of blowing up the Glen Canyon Dam, the five-million cubic-yard plug of cement impounding the Colorado River before it flows into the Grand Canyon, and it features a gang of malcontents who cut down telephone poles, burn highway billboards, and destroy bulldozers on their way to ridding the American West of its military-industrial complexes. *The Monkey Wrench Gang* helped to launch radical, direct-action deep ecology in the United States (which explains the Earth First! logo), and Abbey has long been an inspiration for all those wanting someone more militant than the lyrical John Muir, more uncompromising than the Big Green environmental organizations, like the Nature Conservancy, whose current president, Mark Tercek, worked at Goldman Sachs until the Great Recession of 2008. One of the Conservancy's slogans is "we pursue non-confrontational, pragmatic, market-based solutions to conservation challenges." One of Abby's was "Oppose, resist, subvert, delay until the empire itself begins to fall apart. . . . We will outlive our enemies, and as my good old grandmother used to say, we will live to piss on their graves." It's bracing stuff, and I'll admit to loving Abbey, though it's hard to miss the misogyny, the xenophobia, and the misanthropy that played an ever-increasing role in his writing.

And so the monkey wrench finally also asks us about the role that violence will, inevitably, play in the Anthropocene—the violence of species extinction, habitat destruction, havoc-wreaking weather; the violence of an unchecked chemical assault on human bodies, or the resource scarcity driven by industry and the profit motive. It also asks us about the violence of resistant monkeywrenching, of smashing windows, spiking trees, burning Hummers, pouring corn syrup into the engines of construction equipment, liberating lab animals, blowing up dams. For whose benefit has, does, and will the monkey wrench do what kind of work?

I'm holding my monkey wrench right now, trying hard to hear what it has to say, amazed by the tool's blunt simplicity. What finally occurs to me is that for all its ability there is an awful lot that it cannot do: its reminds me, when I listen close, that the earth is emphatically not in our hands, no matter what the peddlers of Promethean narratives, those who would alter our atmosphere to manage solar radiation, or seed our oceans with iron carbon-absorbing filings, or promise to blast us all off to colonies on Mars, no matter what the technological utopians tell us. The Anthropocene may telegraph the end of Nature, the end of a force always independent from humans, the end of an endlessly exploitable bank of natural resources whose balance can never be overdrawn. The Anthropocene may also be the end of History, as the historian Dipesh Chakrabarty has argued, the end of a distinctly human past plotted against a static, inert natural world. It may be the end of all the old master narratives that have given the modern age its distinctive shape—of the control of nature, of human progress, of a rising tide that lifts all boats—and also of the environmentalists' favored narrative of decline, the one where "man is everywhere a disturbing agent." But perhaps this is a good thing, for the earth, it bears repeating, is not in our hands; only our tools are. And tools are nothing if not the possibilities of a new future made material.

I listen again, and realize that the monkey wrench's greatest strength—indeed, its intended purpose—is to turn the bolts connecting dissimilar things. Perhaps, in the Anthropocene, the wrench has a newfound purpose: securely bolting nature and society—whose separation has long signified the triumph of the modern, capital-hungry world—back together.

I don't particularly like the term "Anthropocene," but perhaps with a little monkeywrenching it can be repurposed. It seems to me that, in the end, the Anthropocene is always a narrative of who "we" are and how we got "here," which is to say that the Anthropocene is always a braided tale of history, people, and place. It also seems to me that wherever "here" is would be better if it was open to, and worked to the benefit of, and was cared for by all the "we" who historically built and continue to build it. There's no great knowledge needed to use a monkey wrench—it's nothing if not democratic—but neither is it capable of intricate work, and there's much beyond its control. It can't alter the past, for instance, though it can fashion a future that finally meets its obligation to justice. And though it can shape the world, it can't control it. Whatever the monkey wrench builds will need to be constantly adjusted to local conditions, and therefore will remain easily manageable by anyone who can hold it.

Maybe this is a version of the Anthropocene that can work: a world—in the making, right now!—for *all* humans; a thoughtful world attuned to the past and to the durable presence of nature, both; a world built to honor that single obligation that Rachel Carson knew bound all living things together, the obligation to endure. Such work is good, whispers my monkey wrench. Such a world is good.

BIBLIOGRAPHY

Abbey, E. 2000. *The Monkey Wrench Gang*. New York: Harper Perennial.
Arendt, H. 1998. *The Human Condition*. 2nd ed. Chicago: University of Chicago Press.
Bookchin, M. 2005. *The Ecology of Freedom: The Emergence and Dissolution of Hierarchy*. Oakland, CA: AK Press.
Solnit, R. 1999. *Savage Dreams: A Journey Into the Landscape Wars of the American West*. Berkeley: University of California Press.
White, R. 1995. "'Are You an Environmentalist or Do You Work for a Living?': Work and Nature." In *Uncommon Ground: Rethinking the Human Place in Nature*, ed. William Cronon, 171–85. New York: W. W. Norton.

The Germantown Calico Quilt

Bethany Wiggin

Close to 1824, at the edge of the North American Atlantic coastal plain, just northwest of Philadelphia, up the long hill to Germantown, Mrs. John Smith stitched a whole-cloth quilt out of cotton calicos. She pieced its reverse from uncut rolls of fabric imprinted with a doubly commemorative image.

On its left side, the print used for the quilt's reverse displayed a half-portrait of the French hero of the American Revolutionary War, Lafayette, depicted in connection with his 1824 triumphal return visit to tour the American republic. It prominently features Philadelphia Quaker landmarks, including the Friends Meeting House on Fourth and Arch streets. On its right side, the print repeats the already well-known iconography of the Quaker city's founding, "Penn's Treaty with the Indians" (plate 10). Mrs. Smith and her quilt connected these most public American republican celebrations with her family's own intimate, private history. She designed and produced the quilt at her home, on the corner of Rittenhouse Street and Germantown Avenue, and today the quilt is part of the collections of Historic Germantown (its fragile condition allows only carefully controlled exhibit). Its accession record documents that Mrs. Smith made it as a gift to her son on the occasion of her own thirty-ninth wedding anniversary.

Mrs. Smith's chosen calico connected her family in Germantown to geographically expansive stories. By 1824 calico's production rested on more than 200 years of global trade in this fabric. Mrs. Smith's selection of this particular pattern was particularly canny, as the print places calico itself at the center of global trade. The pattern in fact compels as much for what it conceals as what it reveals.

Its dizzying *mise-en-abîme* re-enacts the Columbian exchange, recasting Columbus of the black legend as the virtuous English Quaker Penn. It offers us, in fact, the primal scene of the Anthropocene: fast three-masted sailing ships in the upper left hint at the new maritime technologies that moved humans and other animal species, plants, and manufactures across the Atlantic world and across the globe, creating the chains of supply and demand requiring massive resource extraction: wood for ships, and then later, coal, oil, and gas. The image only hints at other technologies that opened markets, by force when necessary: the guns so conspicuously absent in this depiction of the Quaker peaceable kingdom. And it conceals the violence that accompanied cotton's global spread: the slave economy.

The quilted layers recall geologic strata, and it provides an object lesson in the processes visible in the rock, including the Orbis spike of 1610 proposed as the origin of the Anthropocene by Lewis and Maslin (2015). But the Germantown quilt gathers narrative strands for a story of that era that primarily employs synecdoche rather than simile. It offers a microcosm through which we might glimpse Anthropocene macro-structures. Synecdoche, as Hayden White suggested, is an ironic figure, and it best figures the uncanny doublings and reversals that characterize the Anthropocene. To narrate it, we need stories that mirror and reflect through a glass, darkly, oscillating between micro- and macrocosms, and, as this essay discusses first, between places and times, as well as, in its second section, between subjects and objects. These are just some of the Anthropocene ironies that Mrs. Smith's earnest Quaker quilt uncovers. In the Anthropocene we, no less than Mrs. Smith, are caught in a double bind. We are sure of our desire to embody the change we want, all the while knowing we are part of the problem. Our mutual, if uneven, implication in this era is mirrored too in the story of the quilt.

Anthropocene Duplicities

Produced at the corner of Rittenhouse Street and Germantown Avenue, the quilt localizes a term for geological history whose stories mandate a globe frame. The quilt embodies the "glocal"—a word whose cuteness troublingly echoes corporate-ese. The glocal describes a geography in which the local is at once determined by and determinative of the global. In making her quilt Mrs. Smith tied her family into a wider history and a longer historical narrative, one still unfolding. For if the quilt connects Mrs. Smith's familial Germantown history to a whiggish account of American progress, it also ties the American scene firmly to the global, Columbian exchange. English Quaker leader and nobleman William Penn negotiated the fabled 1682 treaty for land rights for European settlers in the newly christened Pennsylvania, Penn's Woods, in exchange for those settlers' continued gifts of consumer goods to Delaware (Lenni Lenape) neighbors. These gifts centrally included yards of East Indian calicos and other fabric stuffs. Global commodities took center stage in this treaty brokered on the banks of the Delaware River.

As an example of Quaker piety, the treaty dealt a very Quaker rebuke to English Puritans in Massachusetts amid the immediate aftermath of the horrors of King Philip's War (also known as Metacom's Rebellion) of 1675–76. But Penn's treaty also served implicitly as an English castigation of Spanish Catholic dealings in the West Indies: conquistadors and crown administrators sullied by their own brutality as well as by the misdeeds of the black legend so eagerly disseminated by presses from Frankfurt to Amsterdam to London for a voracious Protestant reading public. With a reference to the Quaker refusal to swear oaths, Voltaire named it "the only treaty never sworn to and never broken." And while some historians, especially the astute Francis Jennings (2010), have diagnosed Penn as dealing in "the cant of conquest," others, such as the canny Daniel Richter (2013), assert that William Penn really did do a whole lot better by Native peoples than had the Puritans in Massachusetts, Raleigh in Virginia, or the Spanish in the Caribbean and in Central and South America.

The treaty was first translated into a visual medium by painter Benjamin West. Thomas Pierson, West's grandfather, had been present among the Europeans on the Delaware River bank at Shackamaxon, and it is assumed that his eyewitness account informed his grandson's treatment. West's painting drew too on Europeans' ideas about American Indians' natural—that is, unclothed—nobility and rhetorical acumen, long circulating in the cultural imaginary and popularized by Rousseau. As historian James Merrill notes, West's rendition is a "central icon of the American experience" (1999). It came to be so via countless reproductions.

The painting was first copied in an engraving published by Boydell in London in 1775. According to art historian Edith Standen (1964), the copperplates may in fact have been commissioned by Penn's son, colonial proprietor John Penn. Unlike his father, John Penn was decidedly not known for his fair dealings with his Native neighbors. He was one of the engineers of the infamous 1737 "Walking Purchase," which swindled Delawares out of land along the Forks of the river, north of Philadelphia. The Purchase soured Native-settler relations throughout the mid-Atlantic, contributing centrally to Delawares' decision to side initially with the French in the French and Indian War.[1] Though he was no shepherd in a peaceable kingdom, John Penn made extensive use of West's peaceful rhetoric. John Penn ordered, for example, yards of calico imprinted in indigo with the engraved version of West's Treaty. The costly fabric was made into luxurious cotton bed curtains with matching curtains for the bedroom suite's windows.

By the time the up-to-date John Penn received his bespoke order, cotton calicos had been fashionable in Europe for close to a century, gracing well-appointed domestic interiors on wallpapers and upholstery. In addition to their beauty, these printed cotton fabrics possessed an undeniable virtue. Unlike silks or damasks, they were washable.

Original to the Indian subcontinent, calicos had become so popular across Europe by the early eighteenth century that their importation to England, to give but one example, was forbidden in 1701. Such mercantilist policies were common to protect fledgling French, Dutch, and English producers of knock-off chintzes as well as to eliminate competition to European silk and wool manufacturers. Nonetheless, the demand for Indian calico prints far outpaced the ability of European traders to supply it. The demand came not only from Europeans, but also from their American "neighbors" in Penn's Woods and across the hemisphere. As Laura Johnson (2010) meticulously documents, fabric stuffs played an essential role in all British colonial negotiations with Natives in North America; American Indians' pattern preferences are legible in the changes in patterns designed and produced in East India specifically for the American market.

The Anthropocene is the era of slow violence, in Rob Nixon's eponymous phrase, an era of "disasters that are slow moving and long in the making" (2013). Such disasters' creep can be hard to perceive; their toll spans generations and continents. On a local, human scale, they can be difficult to witness (although many American commentators, Native and European alike, frequently noted the devastating losses of Native population to smallpox and other contagions). To make Anthropocene violence legible requires a setting simultaneously local and global, and it urges a historical frame extending at least to 1492.

But the temporality of the Anthropocene is not only slow. It is also fast, and its pace is always accelerating. For the Anthropocene is also the time of global modernity, and it is thus inextricably tied to modernity's temporal regime, that is, fashion's dictate of relentless obsolescence in the inexhaustible quest for novelty. The story of the Anthropocene is thus double both temporally and geographically. Its places are always dislocated, at once local and global; its times are ever out of

PENN'S TREATY.   [1682 or 1683 ?]   Over.

Figure 8. "Penn's Treaty." Ayers Cherry pectoral trade card. Courtesy of Laura Johnson.

joint, both fast and slow. It is doubly double—and often, as the case of John Penn reminds us, nothing short of duplicitous.

Walter Benjamin (2002) espied fashion's palace in the glittering shopping arcades of nineteenth-century Paris, and fashion popularized images such as Penn's Treaty by emblazoning it on consumer goods bought and sold in commercial meccas and colonial outposts across the world. These goods include the fabric Mrs. Smith purchased for her quilt no less than the myriad other consumer goods emblazoned with iconography meant originally, in another ironic turn, to celebrate Quaker plain virtues. Penn's Treaty was thoroughly swept up in the long consumer revolution that began perhaps as early as the fifteenth century (Sarti 2004) and was in full swing by the eighteenth (McKendrick, Brewer, and Plumb 1982).

As Max Weber famously declared of William Penn's grandchildren, actual and spiritual, in *The Protestant Ethic*, "Man is dominated by the making of money, by acquisition as the ultimate purpose of life" (1992). Here too we have another of the uncanny doublings endemic to the Anthropocene: pious practices mirrored and estranged in capitalist culture. The spirit of capitalism forms a major strand in the quilt's Anthropocene story. Objects commemorating Philadelphia's establishment via Quaker plain speech and simple virtues fed supply and demand for ever more consumer goods, including the eerie image in especially poor taste in which calico rolls for the Indian trade are imprinted with advertisements for cough syrup (see fig. 8).

Labor in the Anthropocene

When Mrs. Smith pieced her quilt together about 1824, her cotton would likely not have been imported, although it may have been milled in Britain. The milled textile may have again crossed the Atlantic to have been printed locally at the nearby Germantown Printworks. She made it, as I have already said, as a gift for her son to commemorate her thirty-ninth wedding anniversary. Her choice of print sutured Penn's promise of faithful friendship to her own vow and to her son's future.

In tying them together, she insisted on the unfinished business of the past in the present. William Penn's commitment to the fair treatment of Native Americans rested on his radical Protestant, Quaker egalitarian insistence that the Divine was at home in all living things. It was this insistence that, five short years after Penn's Treaty, saw Quaker Germantown as the site of the first written antislavery protest, authored by German and Dutch Quakers writing in English in 1688. They asked pointedly: "Is there any that would be done or handled at this manner? viz., to be sold or made a slave for all the time of his life?"[2] By 1824 Germantown's antislavery, abolitionist politics had long been established. Anthony Benezet, "father of Atlantic abolitionism" in Maurice Jackson's seminal phrase (2009), printed his first antislavery writing here with the Germantown Saur (or Sower) press in 1759. The Quaker Johnson house, an important hub on the Underground Railway, was only five short blocks north up Germantown Avenue from where Mrs. Smith stitched her quilt at the intersection of Rittenhouse Street.

Four decades before the end of the American Civil War made slavery illegal, Mrs. Smith labored to promote Quaker teachings, including those which had since 1761 prohibited Friends (Quakers) on both sides of the Atlantic from owning slaves. Mrs. Smith's labor, freely employed to make a precious gift, was also a form of religious worship, the feminine, domestic counterpart to the Protestant sacralization of the masculine, professional vocation outlined by Weber. But Mrs. Smith's free labor was all the while haunted by the enslaved labor her Quaker devotion was meant to overcome.

In all likelihood, the cotton for Mrs. Smith's Germantown quilt was produced by slaves laboring in the American South. By 1824, as Sven Beckert (2014) writes, the American South was "in the thrall of cotton." Eli Whitney's gin, invented in 1793, had delivered the key to king cotton's rise to power. And the enslaved labor that produced it provided the linchpin in "the much-vaunted consumer revolution in textiles." By 1802, Beckert notes, "the United States was already the single most important supplier of cotton to the British market, and by 1857 it would produce about as much cotton as China."

Exploring the labor that produced the Germantown quilt illustrates what is the quilt's most important object lesson, one about subjectivity in the Anthropocene. For the *anthropos*, the human, can be either object or subject. To quote Rob Nixon's short essay (2014) on the "promise and pitfalls" of the new era, "We may all be in the Anthropocene but we're not all in it in the same way." The microcosm offered by the quilt is the product, in part, of enslaved labor, labor that is, in yet another turn of the screw, sewn into a quilt meant to be in the service of an emancipatory politics. Read on a macrocosmic scale, the quilt reflects the Anthropocene as at once the time of tolerance and of slavery, an era of human rights and of genocide. Anthropocene ironies abound, rendering linear histories—whether of progress or of declension—a narrative impossibility.

Coda: Utopian Surplus

In his American history, *After Nature* (2015), Jedediah Purdy argues convincingly that the United States' national history "has always been Anthropocene." His argument is grounded in the global dynamism of the Columbian exchange, and as such can be modified for other places too. He writes:

> The human presence in North America has been ecologically revolutionary, wiping out species, changing soils and plant mixes, and reshaping the surface of the earth. At least since Europeans conquered the continent, that ecological revolution has been deeply involved in contests over imagination, over the meaning of the world and the right way to live in it. These are the questions that the Anthropocene finally makes explicit and inescapable: how to live in a world that we cannot help transforming again and again.

Mrs. Smith's Germantown quilt asks precisely this deeply political question, worth repeating, "how to live in a world that we cannot help transforming again and again." I have argued that the quilt provides a synecdoche for the Anthropocene, and so usefully voids teleological narratives, whether they be of rise or fall, of progress or decline.

In *Metahistory* (1973) Hayden White emphasized the impossibility of articulating a political program in a world shot through with irony: "As the basis of a world view, Irony tends to dissolve all belief in the possibility of positive political actions. In its apprehension of the essential folly or absurdity of the human condition, it tends to engender the belief in the 'madness' of civilization itself and to inspire a Mandarin-like disdain for those seeking to grasp the nature of social reality in either science or art." White's claim is true in important ways, and I think it goes some distance in helping us to understand why it is so long and so difficult a path to formulate an effective politics for the Anthropocene: one that would effectively and ethically reduce global CO_2 levels.

And yet for all its synecdocal ironies, I would also claim that Mrs. Smith's Germantown quilt nonetheless prompts deeply political thinking of the sort desired by Purdy, asking the question "how to live in a world that we cannot help transforming again and again." For Mrs. Smith's quilt does not stop in its suggestion that Anthropocene localities are haunted by global relations or that the present time was out of joint. Mrs. Smith's choice of image also insists that there was "something wrong" with her present time and place, "something missing." For this reason, she made her quilt for her son, placing hope in future generations who might explore the utopian surplus of her quilt's meaning.

Mrs. Smith's quilt insists that past business of emancipation and enfranchisement of all was not at all past, that it in fact remained very much unfinished. In this, she recalls, *avant la lettre*, Bertolt Brecht's hero Jimmy in the *Rise and Fall of the City of Mahoganny*. Pressed by his friends to be satisfied in the glittering city, Jimmy maintains, "something's missing." It was, according to utopian philosopher Ernst Bloch in a conversation with his friend Theodor Adorno, "one of the most profound sentences that Brecht ever wrote, and it is in two words. If it is not allowed to be cast in a [finished] picture, then I shall portray it as in the process of being [seen]" (1988). Mrs. Smith no other than Jimmy, refused to articulate positively what the end goal would be of "unbroken faith"; she offered no finished picture. As Adorno said to Bloch, "Utopia is essentially in the determined negation, in the determined negation of that which merely is, and by concretizing itself as something false, it always points at the same time to what should be." Utopia, for Mrs. Smith as for Jimmy, emerges in the opening only revealed by its absence. It is that for which we long.

This is the direction toward which an Anthropocene politics must steer, even as, or especially because, it always eludes grasp. As Bloch noted laconically, "This island does not even exist. But it is not something like nonsense or absolute fancy; rather it is not *yet* in the sense of a possibility; *that* it could be there is we could only do something for it . . . *in that* we travel there the island utopia arises out of the sea of the possible—utopia but with new contents." Mrs. Smith's son, we might imagine, wrapped himself in his mother's gift as he waded out into the sea of the possible. We too could do worse than to find some comfort in its folds.

NOTES

1   The "Walking Purchase" of 1737 played a major role in a 2004 lawsuit brought against the Commonwealth of Pennsylvania by the Delaware Nation (and dismissed). Daniel Gilbert calls it "among the most devastating betrayals ever dealt to the Lenape, the natives who lived on the land taken." Gilbert emphasizes, "Some would call it more diplomatic compared to other white land grabs, and some would say it was civilized for its lack of bloodshed, but ultimately it is remembered as one of the most blatantly underhanded deals ever made by the whites." Gilbert provides maps of the swindling Purchase and references for further reading, including Stephen Harper's *Promised Land* (Bethlehem, PA: Lehigh University Press, 2006). Harper documents how cartographic practices are anything but neutral, "the Pennsylvania proprietors and their agents employed the European weapons of deeds, surveys, and maps to defraud and then dispossess Delawares [Lenape]." And Harper provides a detailed discussion of how doctored deeds and purposely misleading maps were engineered by James Logan and other Penn family associates to convince Nutimus and other Lenape leaders that they would be ceding much less land than in fact became the case. See Daniel Gilbert, "What Ye Indians Call 'Ye Hurry Walk,'" http://pabook2.libraries.psu.edu/palitmap/WalkingPurchase.html (Fall 2009; accessed September 15, 2016).

2   A digital edition of the Germantown Protest, written down by the first mayor of Germantown, German Quaker Francis Daniel Pastorius, as well as explanatory and contextualizing materials are available in the Quakers and Slavery exhibit hosted by the Quaker archives at Haverford and Swarthmore colleges. To access the Protest, enter the exhibit via http://trilogy.brynmawr.edu/speccoll/quakersandslavery/commentary/themes/earlyprotests.php (accessed September 15, 2016).

BIBLIOGRAPHY

Beckert, S. 2014. *Empire of Cotton: A Global History*. New York: Vintage.

Benjamin, W. 2002. *The Arcades Project*. Translated by Howard Eiland and Kevin McLaughlin. Cambridge, MA: Harvard University Press.

Bloch, E., and T. W. Adorno. 1988. "Something's Missing: On the Contradictions of Utopian Longing." In E. Bloch, *The Utopian Function of Art and Literature. Selected Essays*, translated by Jack Zipes and Frank Mecklenburg, 1–17. Cambridge, MA: MIT Press.

Jackson, M. 2009. *Let This Voice Be Heard: Anthony Benezet, Father of Atlantic Abolitionism*. Philadelphia: University of Pennsylvania Press.

Jennings, F. 2010. *The Invasion of America: Indians, Colonialism, and the Cant of Conquest*. Chapel Hill: University of North Carolina Press for the Omohundro Institute.

Johnson, L. E. 2010. "Material Translations: Cloth in Early American Encounters, 1520–1750." PhD diss., University of Delaware.

Lewis, S. L., and M. A. Maslin. 2015. "Defining the Anthropocene." *Nature* 519 (March 12): 171–80.

McKendrick, N., J. Brewer, and J. H. Plumb. 1982. *The Birth of a Consumer Society: The Commercialization of Eighteenth-Century England*. Bloomington: Indiana University Press.

Merrill, J. H. 1999. *Into the American Woods: Negotiations on the Pennsylvania Frontier*. New York: Norton.

Nixon, R. 2013. *Slow Violence and the Environmentalism of the Poor*. Cambridge, MA: Harvard University Press, 2013.

———. 2014. "The Anthropocene: The Promise and Pitfalls of an Epochal Idea." *EdgeEffects* (blog), November 6.

Purdy, J. 2015. *After Nature: A Politics for the Anthropocene*. Cambridge, MA: Harvard University Press.

Richter, D. 2013. *Trade, Land, Power: The Struggle for Eastern North America*. Philadelphia: University of Pennsylvania Press.

Sarti, R. 2004. *Europe at Home: Family and Material Culture, 1500–1800*. Translated by Allan Cameron. New Haven, CT: Yale University Press.

Standen, E. A. 1964. "English Washing Furnitures." *Metropolitan Museum of Art Bulletin* 23 (3): 109–24.

Weber, M. 1992. *The Protestant Ethic and the Spirit of Capitalism*. Translated by Talcott Parsons. New York: Routledge.

White, H. 1973. *Metahistory: The Historical Imagination in Nineteenth-Century Europe*. Baltimore: Johns Hopkins University Press.

Anthropocene Aesthetics

Robert S. Emmett

The Anthropocene travels not only as a scientific and political object of knowledge, but also as a call to artists, writers, and makers of many kinds to reflect on human un-making of living worlds. Confronting the term necessarily involves aesthetics—the realm of sensation, feeling, and apprehension that subsides into muddled emotions or quickens to thought. In collecting for a Cabinet of Curiosities, we open a categorization of wonders that belongs to an era before modern museums, when natural objects (bone, antler, coral, shell) might sit beside products of human artifice (engravings, sculptures, miniature landscape paintings). As this new term splits deep time, wrangling natural and human history together again, more and more writers reach for swift metaphors to abbreviate its epochal significance: awareness of the Anthropocene is like a new world order or a Great Awakening. Jared Farmer explores one of the most hopeful versions of Anthropocene thinking in this section under the metaphor of a "planet of art," an earth where recognizing the ubiquitous effects of human activity might lead to more intentional, playful, or beautiful redesign. The Anthropocene term has certainly inspired a multitude of contemporary and art-activist responses. This collection joins, too, in the polyphony of Anthropocene poetries and its inhuman music and stillness—the echoes of extinct bird songs that Julianne Lutz Warren describes, the silent glimpse of sea angels through a microscope by an imaginary polar explorer in Judit Hersko's dreamlike narrative.

Artists and makers responding to the Anthropocene idea are freely revisiting the technological sublime, articulating a new proletarian literature of transition in the era of Tough Oil (e.g., the graphic novel *I.D.P. 2043*), and designing apps to engage the estimated 2 billion smartphone users in reading land use, real estate, and rising seas. Imagination fires in the gap that opens between perception, reason, and values as we encounter as an overwhelming totality the evidence of human planetary effects. These now include an encyclopedia of circumstances not traditionally indexed under the sublime or beautiful: mining pits that eat towns, oiled seabirds and deformed amphibians that wash up on tourist beaches, children's drawings that inadvertently depict their own incremental poisoning. The dominant political and economic entities for much of the colonial and industrial era (the "short" Anthropocene, or as Sloterdijk proposes, the Eurocene) arguably produced Eurocentric aesthetic forms, the novels, sonnets, and marble torsos that eventually displaced the oddities of the aristocratic *Wunderkammer* in the emerging hierarchies of high culture.

Such objects will continue to be treasured and insured against rising seas in a new Europe, although they are not the only Anthropocene objects worth cherishing. Other objects are embedded in new stories of more-than-human meaning, stories that might unlock resistance to neocolonial objectification, signal a more global estrangement from business as usual, memorialize plural beauties, open pools of deeper reflection onto or perhaps refract the mesmerizing potential of the Anthropocene-as-mirror.

Many have objected to how techno-optimism and unacknowledged normative agendas are being slipped into Anthropocene-spun narratives. The observation that humans have already, if unwittingly, modified earth systems has been put to service to justify further large-scale projects. The Anthropocene, in other words, can be repackaged to favor developmentalist ideology, a "Big Dam Theory of Global Eco-modernity," epitomized by national megadam projects for hydropower, irrigation, and flood control—with human and ecological costs. The ideology extends to less publicized projects, as captured in the painting *Davies Creek Road* by Trisha Carroll and Mandy Martin in this section. The storied landscape in Carroll and Martin's canvas, layered over with the figure of the goanna lizard in X-ray style, offers texture and meaning where the Australian government sees only a blank slate for a proposed dam. Before the Anthropocene becomes a single perspective, story, or agenda, it can still be used to name a raft of forces that resist a simple ending.

The Anthropocene does not necessarily imply a destination—several have pointed out the paradox that it's the first epoch that can begin, but can never end. So can there be no avant-garde of Anthropocene art if we are on a road to nowhere, running in circles? Is there anyone left to shock? Judging from the tens of thousands of visitors to the Deutsches Museum exhibit on the Anthropocene that opened in December 2014, there are, it turns out, many people shocked to discover the extent of our activities' impact on planetary biophysical systems. We need emotionally powerful works of art that reorganize our structures of feeling around these transformations in environment and society.

The Anthropocene framing as a normative discourse is embedded in what Raymond Williams eloquently called "structures of feeling," a term that embraces both particular works of art, genres, and accompanying practices. The Anthropocene framing often comes prepackaged in aesthetic forms, including the manifestos of political think-tanks and dramatic monologues of geophysicists, with a less than imaginative range of feelings. Dipesh Chakrabarty (2012) intimates this in the closing passages of "The Climate of History" and explores elsewhere how an awareness of the conjuncture of geological, biological, and human history poses a challenge to imagination. This is a ramping up of what literary critic Lawrence Buell called for in the late 1990s—a richer, more plural environmental imagination for an endangered world. It was also anticipated by Ursula Heise's articulation of the need for critical attention to problems of scale and planetary imagination in *Sense of Place, Sense of Planet*. The work of imagining the meaning of phenomena now gathered under the Anthropocene banner has already been undertaken by multitudes of individual artists, writers, activists, academics and professional organizations, NGOs and government cultural and ecological or natural resources agencies.

At the very least, artistic experiment and creative exercise is necessary to answer questions raised by the Anthropocene's key insight into the ubiquitous and unpredictable dereliction of human activities on this planet. What might a livable future look like, fleshed out beyond the spidery lines of the Intergovernmental Panel on Climate Change's projection curves? Many of the fictional scenarios that accompany such projections are not survivable for most of us—especially when *us* includes the welter of nonhuman life that we are, too. Even if it were only Western bourgeois *human* livability we were after—an easy chair, a good lamp, a cozy room of our own with a locked door—we would need to consult science fiction novelists and architects as well as biologists and geophysicists.

Each of us might construct Anthropocene Cabinets of Curiosities—or perhaps do so in communities as "little free libraries," where the libraries also contain seeds, specimens, and directions for reanimating forms of extinct life (see Hennessey in this volume). What if the Cabinet of Curiosities for the Anthropocene were also an aesthetic survival kit, a potent dream of a shareable planetary society that prevented numbness to loss? As Daegan Miller pointedly has asked: who will do the work of constructing such a future? And why do this work in miniature, even in the memorable form of these time arks, these Cabinets of Anthropocene curiosity? One answer is provided by Wendy Wheeler when she describes the life of "new semiotic objects": ideas and things, but also poems, paintings, and works of art that "are not doctrinal advertisements for the promotion of whatever local virtues are currently approved; rather, they constitute relationships and serious demands for the light of other minds and room to grow" (Wheeler 2014, 78). We need such objects, the light of other minds and room to grow, because we stand at a point where governments seem criminally unresponsive to the red lines of climate change and scholars are seriously debating whether past human actions necessarily imperil Enlightenment ideals of human freedom.

Works lose their audiences, but forms also age and lose their potency. Nineteenth-century novels documented how the mansions of modern freedom also had their hinterland victims in overseas plantations—hence, the "Plantation-ocene." This connection is taken to heart by Bethany Wiggin's meditations on a bright past future: the Peaceable Kingdom motif of William Penn's peace with the Lenni Lenape. The image was woven into domestic textiles emblematic of self-reliance, respectability, even domestic contentment; it also functions as a warning and a reminder of opportunities missed. The highly textured, home-spun, built-to-last textile is a precursor of DIY and maker culture, the new media ecology of swapping design techniques for the future fossils that Jared Farmer describes in "Technofossil." This exchange of ideas is something more ambig-uous than revolutionary and liberating. Sverker Sörlin opens the prose poem on the mirror as icon for the Anthropocene with precisely this juxtaposition: "The Revolution will not be televised. / The Anthropocene is still on Twitter." One form is not enough: even the mirror develops multiple exposures. Plural human figures pass through its field of vision rather than a singular, narcissistic fixation.

Historian Julia Adeney Thomas, responding to Chakrabarty, has described the human figure of the Anthropocene according to climatologists and geosci-entists as a metaphor that would singularize a plural biology. Is the purpose of metaphor to take the measure of "the human" to better govern (read: dominate) diverse populations? Microbiologists, she points out, describe us instead "as a coral reef of multiple species," and biochemists "examine the industrial toxins suddenly infiltrating our bodies, including our brains," so that, at a neurological level, we are not justified in positing a stable, universal "human" response to the world (Thomas 2014, 1592). Literature and the arts confirm the rich plurality of responses to toxic environments, including fictional and filmic representations of ambiguous ecosicknesses described by critic Heather Houser (2014)—or as a kind of "bad love" in the slippery ways oil enters and makes our lives into their energetic, mobile, lively forms, as illuminated by environmental cultural studies scholar Stephanie Lemenager (2014, 11). Surely thousands of other objects of energy belong in Anthropocene cabinets outside of the US-Western Europe axis of living oil: cosmologies as equally terrifying as those presumed in the unworlding force of the *Plowshare* promotional film that Joseph Masco sifts from the remains of the Anthropocene.

The lively pluralism of the stuff of lives and their multivalent energies resists a singular, totalizing aesthetic of the human in the Anthropocene. That version of Anthropocene aesthetics is crudely reminiscent of totalitarian public sculpture: out marches the human figure again, a muscular solitary male, hero of a new age. Judit Hersko's fictional probing of the gender politics of polar exploration unveils such solitary figures of the human as quixotic or just plain fictions. Similarly, science fiction and the wider category of speculative fiction have mined the history of past futures to remind us of forgotten alternatives. Thinking that we belong to a new epoch can cause vertigo and tunnel vision. The basement of history suddenly falls out, the concrete foundation liquefying. As we accelerate further, are we falling or flying into this late phase of the Anthropocene? Its secular jostling itself fits into a sequence of hefty terms: modernity, postmodernism, globalization, now Anthropocene. Like those earlier discourses of rupture, it is also a foil for intergenerational dialogues about what constitutes good art (or good politics). There are YouTube channels and Flickr groups for Anthropocene, but one can also imagine a commissioned Anthropocene symphony, played perhaps less often than online playlists where Anthropocene is just one more metadata tag. Cultural theorist and poet Joshua Clover launched #misanthropocene into the Twittersphere already some years ago; perhaps the Revolution *will* be on Twitter. Anthropocene frames modernity again; it might also call for different proportions to our perceptions and pose seemingly mundane questions: how ancient and how permanent are the effects of our agriculture, cities, and power plants? Anthropocene can produce a spell of absorption that oscillates between abstract thought and humble matter. Sörlin looking into the mirror as prime object of Anthropocene thinking asks, "Where is the Anthropocene Society?" This can be taken as a call for fresh thinking about humanity, to ask what are these humans in this haphazard age of humans? Or, the Anthropocene mirror test might remind us to laugh playfully at human narcissism: is *this* the face that launched a thousand #misanthropocene followers?

Turned in a different light, the mirrored surface could bring our attention to the social geography of differences in other materials. Another collective gives concrete a voice like a tragic chorus in this volume: concretes speak from rotting sarcophagi over spent nuclear reactors, from beneath overheated seas sculpted into superstructure for cultivating coral, poured in pylons to elevate flood-prone housing in Suriname, Vietnam, and the Netherlands; from tagged cement under crumbling bridges in Vilnius, Milwaukee, and Rio. Concretes speak to an Anthropocene aesthetics that ripples out in collective projects of infrastructure for life.

I have been dealing more or less formally with Anthropocene aesthetics, but hearing in the voice of concrete how the Anthropocene converges with the promise of innovation in a STEAM agenda begs the question: where are hard work and craft patronized under the new epochal sign?[1] From the perspective of a sociology of art, epitomized by Pierre Bourdieu's work on fields of cultural production, artists and writers vie with one another by distinguishing their work through new styles, formal experimentation, and networks of prestige. The Anthropocene also makes art happen in this way. When we use it, we deploy a boundary term in a field of cultural production—its universal reach is not part of a universal literacy. In performing the trope of rupture—"calling into question all the old certainties," as the literary modernists might have said—the Anthropocene also pays bills, opens exhibitions, and feeds new grant programs.

This secondary awareness of how the *idea* of the Anthropocene creates new relations of meaning and intellectual value among forms of knowledge and cultural institutions might seem cynical, but it cannot be dodged. Call it the social epistemology of Anthropocene arts and culture, which holds even in the absence of an Anthropocene Society. Hopefully the conceptual container of the Cabinet of Curiosities functions more like a vampire meme than a genre of fixed social value. The meme catches hold of imaginations as people outside of museums and universities pitch new objects from their lives, talk back to the Anthropocene as they have already done in Madison, Wisconsin; Zurich, Switzerland; and Sydney, Australia. One of the questions we posed in the museum installation in Munich was simple: where should the Cabinet of Curiosities go next? The Cabinet can be an experimental record and a template.

We collected objects with the quality of ubiquity, curious physical appearance, as well as their individual resistance to meanings attributed to the Anthropocene. This play of resistance and discovery could be molded into an object's qualities—for example, the wisp of an unknown bird feather in Gary Kroll's "Snarge" sample; or the transparent, rubbery surface of Hersko's silicon print, the surprise of how it reveals sea snails and a portrait of Anna. Or it could come primarily from the resounding cultural symbolism of an object such as the monkey wrench. Biosemiotician Timo Maran has usefully described as the "semiotization of matter" this complex interplay of meanings that objects provoke and the stories that sometimes seem to automatically unspool when we contemplate a pipe, a dam, or a plastic cup (2014, 143). As the human world has come to be co-extensive with the entire planet, there is no object now on Earth that is not already somehow an "Anthropocene object" in this general and damaged anesthetic sense. In this way, too, the Anthropocene nudges us back to the modernist poetics of the fragment, further back to the Romantic aesthetics of ruins, even as it runs forward into declinist and futurist studies.

The objects collected and scrutinized in this book reflect a search for useful insights into environmental futures, plural, but also a conviction that whimsy, wonder, and the unexpected are often welcome emotional and intellectual guides. Some of the policy prescriptions and ideas called for in public discussions of the Anthropocene resemble triumphalist narratives of one world system, one official politics, and one sanctioned form of art. The mirror, the future fossil, Anna's portrait, and *Davies Creek Road* steer the conversation in different directions. How we make a better environmental future from the predicaments of being just humans calls for the talents of historians, storytellers, artists, critics, anthropologists, biologists, and the creative energies of many others.

NOTES

1   STEAM adds Art and design to the more familiar STEM (Science, Technology, Engineering, and Mathematics) and argues that innovation results from imagination, creativity, and the qualities nurtured by the fine arts: http://stemtosteam.org/.

BIBLIOGRAPHY

Chakrabarty, D. 2012. "Postcolonial Studies and the Challenge of Climate Change." *New Literary History* 43 (1): 8–10.

Houser, H. 2014. *Ecosickness in Contemporary U.S. Fiction: Environment and Affect.* New York: Columbia University Press.

Lemenager, S. 2014. *Living Oil: Petroleum Culture in the American Century.* Oxford: Oxford University Press.

Maran, T. 2014. "Semiotization of Matter." In *Material Ecocriticism*, ed. Serenella Iovino and Serpil Oppermann, 141–54. Bloomington: Indiana University Press.

Thomas, J. A. 2014. "History and Biology in the Anthropocene: Problems of Scale, Problems of Value." *American Historical Review* (December): 1592.

Wheeler, W. 2014. "Natural Play, Natural Metaphor, and Natural Stories: Biosemiotic Realism." In *Material Ecocriticism*, ed. Serenella Iovino and Serpil Oppermann, 67–79. Bloomington: Indiana University Press.

Making

The Mirror—Testing the Counter-Anthropocene

Sverker Sörlin

The mirror is the test of a reflexive Counter-Anthropocene.

The Revolution will not be televised.

The Anthropocene is still on Twitter.

Anthropos means human. Cene means new. What is new?

Where is the Anthropocene society? Or is "society" just something else, alongside the strata?

Should we now stop thinking about Industrial society? Knowledge society? Capitalist society? Of "New heavens and a new earth"?

Or is the Anthropocene precisely a "New heavens and a new earth"?

The mirror was the selfie of the ancients.

Do I in this mirror here see the face of God?

How many faces of God can we imagine?

Which ones of these 7.3 billion Gods drive cars? Or have voting rights?

The Mirror is a test of hope.

Anthropocene is high risk.

Risk is the twin sister of progress, the twin brother of profit.

A mirror is the entry to Wonderland, says Alice.

In 2005 it was proved that humans move around more gravel, soil and stone every year than the Earth itself.

"The Earth will never escape the hands of man," said Italian geologist Antonio Stoppani in 1873.

In 1922, in Paris, Lucien Febvre, published *La terre et l'évolution humaine*, an attack on determinism. Environment gave possibilities, and demanded responsibility.

In 1923 Polish scientist Antoni Dobrowalski published a book about the Cryosphere, *Historja Naturalna Lodu* (Natural history of ice), and how humans could melt ice on a grand scale.

Vladimir Vernadsky invoked in 1924 the psychozoic era of human involvement in chemical circulation of agriculture and infrastructures. Man was a *Homo sapiens faber*. He also proposed the Biosphere.

Anthropocene started with a human face—soon a hundred years ago.

The Anthropocene is political.

Political is a word we rarely read in *Science* or *Nature*. The *Economist* is more honest.

The Stratigraphic Commission will not decide on justice; living together is not yet defined as a geological problem.

The next experts of the Anthropocene must also be experts of the Anthropos.

Parts of the world will be turned into deserts. Who will decide on who must live there? The market?

It should be called The Econocene, says Richard Norgaard, the Berkeley economist.

It is the thinking of our time: to marginalize common human judgment.

It is a time of numbers. Planetary boundaries, determined by science.

The master narrative is about *global* change. Most people will soon know the story, just as we know about the atom.

There is no master narrative about social change: We arrived here inadvertently. There was no roadmap.

Or perhaps there was. There are people who believe that the Anthropocene is a version of human fulfillment. Manifest destiny.

There is no backdoor for the doubting.

The Anthropocene is a case of forced collectivization.

The naked beggar, the outcast, and the starving child all live in the Anthropocene.

Fifty years ago Frantz Fanon said, as Sartre put it in his foreword to Fanon's book, that the rich and white were vulnerable, illuminated paper faces hovering in the dark.

The wretched of the earth would come after them. Or us.

Now we are all equals in the face of the planetary boundaries.

We are all part of the species.

It is like war. A general conscription. If it is planetary—there is no refuge.

Still—we disagree about climate change. We even disagree about creation. We of course disagree about the end of the earth.

Few disagree about the virtues of capitalism.

It is said that we are running geological time backward. We are creating a new Pangea.

Oh, this great extraction of fossil fuels that we are all collectively engaged in.

Are we?

Am I?

Dare I see myself in the mirror?

There is a "so what?" question of the Anthropocene that remains unanswered.

Unless you take the US Congress for an answer.

The mirror is a choice.

Of surface, of now and just now.

Of what is underneath, how we became us, how we became insides, too. How we became divided already in the Pleistocene.

Reawakenings—to the questions of time.

Humans see themselves. One human sees herself, or himself.

The environment is human. This is what we have learned.

What can you see beyond the mirror?

How can you see beyond surfaces?

What is inside us unites us. Tissue is the largest democracy.

We share in toxics. Our organs swim in pain, in pleasure.

Narcissus faces a pond of water which connects humans with the chemical world of our own making.

Thoreau's words come to mind: "I am nature looking into nature."

The Anthropocene is about the now. If we wish it to be about the past and the future we must return to our insides. The mirror will not let you know.

We will soon agree on the Anthropocene. Courses will be renamed—"From the Atom to the Anthropocene." Lecture Halls will be called Vernadsky and Crutzen, hopefully also Stoermer.

Will they be called Fanon? Febvre?

Why are there no women in our genealogies? Our histories must be skewed.

That women take the lead in Anthropocene critique is probably a sign.

Caution. Caution.

We disagree about social change. We disagree about who we are and who we ought to be.

We disagree about the very word. "We." We are all in this together, we are all born equal, but we are not all equally responsible and we do not all fare equally well.
Who do you see in that mirror? Are you one of the Gods? Belonging to the God species?
Then, what do Gods do these days? How do they go about their business?
And—why does your face look so tired?

Twenty five years ago today the Berlin Wall came down. The people did not allow the Wall to become part of the Anthropocene.
When the Great Acceleration started there was a lone refugee sitting in Harvard's Widener Library writing a big book. His name was Ernst Bloch. The book he wrote was *Das Prinzip Hoffnung*.
The principle of hope. He took it to East Germany. It was printed there in the early years of the Anthropocene. It grew tacitly—under the surface of a wrongful society. Ideas can make walls come down, unsettle the Anthropocene.
We are at the beginning.

Testing the Counter-Anthropocene

> Could it be that democracy is such a hit
> with modern humans precisely because it
> mirrors our greatest folly—our nearsightedness?
>   Arundhati Roy

"The mirror test" is a check of whether you have an idea of who you are or, perhaps, that you are at all. Elephants and apes pass this test, as do also, we have learned, the magpie, and humans from the age of eighteen months. Seeing ourselves in the Anthropocene mirror we stand a slightly different test. Not only: do I realize that I am there? But: do I realize that I am part of something larger? Do I figure what this larger something might be?

The Anthropocene mirror as a displayed item in the Deutsches Museum's *Willkommen in die Anthropzän* exhibition appears in dual form. First, as a physical mirror where everyone could see himself, an everyday experience, but one that becomes different when it happens in an exhibition—where you are also transformed into a displayed object, sensing the awe and responsibility of being enhanced to an agency observed not only by yourself but also by others (plate 11). Thus you are literally becoming part of the Anthropocene, facing a moral, and ultimately political issue: *Am I a part of all this? What will I do about that?*

Second, the Anthropocene mirror *Poem* is accompanied by a *Video* presenting a mirror as human comedy, showing a few members, a small fragment of the collective *Anthropos* that the Anthropocene presupposes. These *anthropoi* stand behind this object, acting it—and also in front of it, in a very literal sense. A group of undirected environmental humanities scholars, PhD students, and some of their supporting staff in Stockholm, located not far from the melting glaciers whose Mother Ice some 15,000 years ago was three kilometers thick precisely where this little herd of humans have now built an Environmental Humanities Laboratory, where they ask questions such as these and debate the Anthropocene intensely.

Our choice of the mirror as the metaphor, and as an engaging object, is predicated on what we perceive as a humanity that is positioned in many different realities, and stands in multiple relationships with the Anthropocene. But a humanity that also stands in multiple, potentially an endless number of relations to the world the Anthropocene assumes. This is not just *one* world where a separate humanity impacts on everything nonhuman but a world of increasing entanglements across scales and species and forms of being in the world and thus a world of multiple becomings. As a sum total, humanity may still satisfy the equations of the Anthropocene, but every single encounter is different and each and every one of us has a personal responsibility, which is embedded in our histories (Golub, Mahoney, and Harlow 2013).

To display the mirror is in and of itself an act of inclusion, despite the seemingly private and intimate interaction it symbolizes. The mirror once was an object of luxury, something you would find in an emperor's Cabinet of Curiosities, or in his grave. But the mirror has become universal and nowadays can be owned by everyone, part of consumerism and itself intertwined with the forces that forge the Anthropocene. Everyone has a face, which, as already Emanuel Levinas insisted, is a moral instance: when I see someone else in the face I have to take responsibility for my action, or inaction, for who I wish to be and be regarded as. Seeing your own face makes no difference: am I whom I wish to be?

Relativizing Human Hegemony

The mirror is a border, a utopian line. It raises issues of belonging. Is it just me? Or are there more of us? Who are the mirrored ones? These questions point toward the mirror as also a critical instrument: part of a method of critique that allows us to extend criticism beyond the work of humans.

The Anthropocene discussion has already started to serve as a mirror, inviting us to see our world, and our position in it, more clearly and earnestly. Some recent work in fields such as philosophy and ecocriticism is fusing ideas and concepts from posthumanist discourse with the emerging image of humanity as a hegemonic and mostly destructive external agency. From this work emerges a way out of the seemingly confrontational attitudes that have taken root when environmentalists have considered the human-centered, sometimes triumphalist language of the most ardent defenders of the "good" Anthropocene among self-professed representatives of the "God species" (Lynas 2011). This skepticism of the Anthropocene is also an articulated presence among many scholars in the Environmental Social Sciences and Humanities (ESSH) (Castree et al 2014). There are certainly good reasons to be skeptical of an Anthropocene discourse that seems only too oblivious of the complexities of a 'humanity' that is not as singular and unified as the mainstream discourse presupposes (e.g., Pálsson et al. 2013; Crist 2013; Malm and Hornborg 2014). But is it the only way of looking at this concept?

A counter-Anthropocene understanding is emerging that takes the earth systems science's factual and analytical approaches to the new era onboard, while at the same time building tirelessly on its wider *Weltanschauung* and making explicit what it sees as a necessary ethics and politics. From such a closer look at the interface between the human and the nonhuman features of the Anthropocene there have emerged those new and more-than-human properties and agencies that change the interface itself and make it more permeable. If examined more carefully, "humanity" isn't appropriately described as an external megaforce reshaping the world almost on a biblical scale. Nor is it the world that transforms us. Rather it is both at the same time. Humans are *of* the world, argues Joanna Zylinska, and are just one of any kind of entities in the world (2014). Biosemiotics has pointed to deep similarities and affinities between the way elementary life forms unite humans with the natural world (Ingold and Pálsson 2013) and is now being extended to the nonliving, as it were "purely" geological realm as geosocial becomings (Pálsson and Swanson 2016). Indeed, the very concept of becoming (Connolly 2011) suggests new approaches to time and temporalities and alternative ways of looking at causalities and connections uniting subjective time-perceptions with objective human-ecological processes on quite different time scales. If so, becoming would be moving along lines that were not attainable to Henri Bergson, writing on subjective time before environmental agency in humans and the agency of matter on the human were conceivable.

Still, this may seem paradoxical insofar as it is precisely the external human hegemony, the cornerstone feature of the Anthropocene, that is being questioned by an agency that in this counter-Anthropocene is attributed to the posthuman

(Braidotti 2013; Barad 2007) and to matter itself (Bennett 2010). The Anthropos of the Anthropocene—the individual, autonomous, quintessentially human of the ancients and of classical humanism—may in fact be a posthuman feature. Seeing ourselves in the mirror will thus not only be to see the faces we are used to but another kind of human being, with far less clear boundaries to the rest of the world, and permeated by substances from the toxic environments that humans themselves made toxic in the first place (Alaimo 2010). Some of the most vanguard work in critical environmental humanities of art and literature explores precisely this intersection of humans and the post-human. As Serpil Opperman (2013) has suggested, there is a "diffractive" ecocriticism that presents itself in relation to the creative nonhuman part of the world.

Another major paradox lies in the fact that the very age of the Anthropocene—when the post-1945 Great Acceleration multiplied human resource use and earthly impacts—was also the period when humanity became increasingly aware of its role without being able to do anything about it (Stoner and Melathopoulos 2015). This Anthropocene helplessness is a remarkable flip side of the triumphalist progress of mankind. Precisely the hegemony that Anthropocene presupposes is being undone by this defeat of the capacity to pursue collective action. Strangely noncollective action—myriads of ill-conceived, individualist, if not egotist, actions—lead together as by some invisible hand to the massive blow to creation. This reflection alone should caution us not to take human superiority too seriously. Anthropocene could be seen as the ultimate weakness of humanity, its demonstrable failure to get itself organized to defend what is important over the long term and in its own self-interest.

The mirror test thus works on humans as well. Do we see our own fate in our faces? Can we work up that probing, examining, yet compassionate gaze for the feature that we used to regard as "us"? In the kinds of literature in the ESSH and the arts that I have just cited we can identify "mirroring" almost as a method of Anthropocene ethics and a technology for questioning the narrowly narcissistic approaches that hold up the concept merely as the most recent stage in a sequence of layers. Levinas's ethics of the face didn't originally include the entanglements and becomings of the current posthumanist discourse but now is gradually absorbing them.

Justice and Temporalities

The heated Anthropocene debates are so heated at least partly because of the growing dissatisfaction with the scientist, or STEM (science, technology, engineering, mathematics) hegemony over the environmental expertise, and the buildup of environmental social sciences and humanities. The ESSH have been enormously stimulated and aided by the rise of civic environmental movements and NGOs and the ever more forceful organization of environmental justice issues (Armiero and Sedrez 2014), self-consciously launched under this name since the early 1990s (Bullard 1990; Martínez Alier 2002) but with roots in many diverse and local grassroots protests in countries all over the world and often linked to race (Finney 2014), class (Hurley 1995; Nixon 2011), gender (Merchant 1980, 1996), and not least indigenous politics (Guha 1989).

The environmental justice and political ecology literature also provide an essential local counterpoint to global-scale narratives. This scholarship has often been critical of the Anthropocene approach, arguing that "the age of humans" obliterates the dramatic inequalities among human beings and assumes a "we" that is difficult to recognize in the real world. Environmental justice scholars and activists have expressed scepticism toward discourses premised upon transgenerational equality or future apocalypses, arguing that these are contemporary and not future problems (Swyngedouw 2010). Although the environmental justice literature stresses the challenges of the present, understanding the past is essential for recognizing and resisting the naturalization of inequalities in the present (Agarwal and Narain 1991; Davis 2001).

The temporalities in the environmental justice literature are rarely those of deep geological time; they are always historically and spatially situated. The focus on power relations is crucial for bringing social relevance to debates over the starting point of the Anthropocene (Industrial Revolution? Atomic age? The Neolithic revolution?) (Frawley and McCalman 2014). There have been efforts to try to compare the ecological footprint of nations, while other scientists have tried to complicate the exchange relations between rich and poor countries by including measures of ecological change (Hornborg 1998; Røpke 2001).

While climatologists and geologists play visible roles in the mainstream Anthropocene literature, the environmental justice reading of the Anthropocene foregrounds ecologists, energy scholars, agronomists, and medical professionals. Public health and epidemiological studies are central to analyses of the unequal distribution of risks and environmental burdens. The political invisibility of slow violence can be related to the longer timeframes needed to determine with certainty the effects of human activities (from carbon emissions to pollutants) upon environmental systems. Forms of knowledge that identify these processes at local scales before the stamp of scientific certainty can be applied—often well after the changes have wrought damage—are therefore receiving greater scholarly attention (Corburn 2005). The traces of the Anthropocene may thus be found in and on bodies in addition to the strata analyzed by geologists (Guthman and Mansfield 2013; Mansfield, 2012).

In the wider scheme of things the Anthropocene debates are therefore linked to an ongoing dissolution of the linear, progressive time in the study of modernization theories (Koselleck 2000) and the cultural studies of memory (Nora 1989; Coser 1992; Halbwachs 1925). Science-informed narratives can be seen in this context of destabilizing of a universal history. The French historian Francois Hartog (2003, 2013) has used the concept of *regime d'historicité* to argue for a breakdown of the modern "temporal regime," symbolized by the year 1789 and the arrival of the future as the hegemonic dimension of time. Hartog's "crisis of time" since 1989 signals an increasing conflation of past and future into a widening "present" in which temporalities lose their immediate sense of beginnings, ends, rises, falls, and ultimately directionalities. For this age of growing contemporaneity, or "presentist regime" in Hartog's terms, to become real we must assume a set of historiographical avenues with which to establish this expanding synchronicity. These "practices of synchronization" (Jordheim 2014) provide a useful theoretical concept to organize the empirical enterprise of locating and researching the actual social, intellectual, political, and other processes that work to establish current historical outlooks.

An Anthropocene of Hope

In such a reading the Anthropocene discourse—from Paul Crutzen's first oral use of the term in 2000 to the dictum of the Stratigraphic Committee of the Royal Geological Society in 2016—stands out as a cardinal example of a "practice of synchronization"—an attempt to take control over time and thus through the immense cognitive and emotional and symbolic powers of temporalization extend this control to the political sphere of action and decision-making. If this is only marginally true, the choice of metaphor for talking about this practice comes across as no innocent thing.

The metaphysical linearity that is often taken as a given in the Anthropocene narrative, especially with its inclusion as the latest (last?) of all stages in the evolution of Earth and all of life upon it, is an understanding that, if left to itself as a mere geological fact, is problematic at best. First of all, in this version it gives undue singular power to an idea that is much older than the recent Anthropocene flare up since 2000; the roots date back at least to Buffon in the eighteenth century (Zalasiewicz et al. 2018). The formation and ascendance of "the environment"since World War II likewise already contains many of the essential ideas that the Anthropocene now is proposed to bring (Robin, Sörlin, and Warde 2013). Second, it runs the risk of reverting to a narrowly science-based interpretation of relevant knowledge by focusing on the geophysical strata and the refinement of the science narrative, at the cost of the social complexities that both lie behind the massive human impact and will undoubtedly follow from it.

Linear Anthropocene also makes it hard, if not impossible, to perceive the full range of forces at play. Thus the linear, or perhaps more precisely sequential, understanding of the Anthropocene serves in reality as a political device that obscures some of the other properties of the epoch that humanity has been entering since the start of the Great Acceleration: that it is a period of growing inequalities, enormous accumulation of wealth in the hands of a few, an emerging risk society, a new geopolitics of natural resources and energy, and massive transitions across societies and entire regions, such as the Arctic (Wormbs 2013)—some undoubtedly for the better, others surely for the worse.

This is not an innocent thing, because words are no innocent things. Concepts, as a host of social thinkers from Foucault and Koselleck to Latour have taught us, have power. To establish the Anthropocene as predominantly a geological fact, a naturalized and inevitable moment, or plateau, in a metaphysically understood evolution of human-nature relationships, verges on stripping the human condition of its largely benign search for new directions and to argue that certain properties of the world are more respectable than others. After all, there is nothing metaphysically linear or unidirectional in the sweeping changes that are shaking the planet and its growing population. The often-used image of a sedative humanity, overdosed and greedy and collectively incompetent to deal with its predicament is not valid. Humanity is divided, between nations but also within nations (Jamieson 2014). Some are greedy, some are inert, some are robbed of common sense by their material success. But most people are none of those.

That is why what we need now is to move ourselves into center stage not just as a collective species agency, but as a moral subject. The future of the Anthropocene will be dependent on whether humans and their societies can move on, from the facts and descriptions provided by earth systems science and geology, to the (self-)critical reflexivity whereby we ask ourselves and our fellow citizens how we can and should relate to the Anthropocene predicament. The poem gestures at looking into the mirror as a moral act, as it has always been: a moment of potential self-examination. It is also a political act, a moment to reflect on how we still carry responsibility in the world and for the world. It *is* a test: can I see myself going beyond the present state and induce hope for a better state of the world? Can I even think of the Anthropocene as a *mere* geophysical fact, the "signals" of which will stay forever in the earthly strata but which shouldn't stop us from leading our future presence on the planet in new directions.

Once you have scrutinized yourself, the mirror you hold could be used in so many other ways. Mirrors are in telescopes and microscopes, in cameras and periscopes that help us see back and forward, sideways, above the surface, beyond the stars, and around the corner. Put together they could make us see ourselves from any angle. That a human being is there, seeing, doesn't exclude that what you can learn to see is a world of becoming and entanglement—an Anthropocene of hope.

NOTES

"The Mirror—Testing the Counter-Anthropocene" is a poem, recital, and video installation with an accompanying essay. It was originally presented at the Anthropocene Slam, University of Wisconsin, Madison, November 9, 2014, by Sverker Sörlin.

Poem by Sverker Sörlin with inspirational support by from Marco Armiero.

Video installation produced by Johan Gärdebo.

Original mirror idea by Marco Armiero elaborated further by Anna Svensson and other scholars, students and supporting staff at the Division of History of Science, Technology and Environment, and the KTH Environmental Humanities Laboratory, KTH Royal Institute of Technology, Stockholm.

Accompanying essay by Sverker Sörlin with contributions from Marco Armiero and Nina Wormbs.

BIBLIOGRAPHY

Agarwal, A., and S. Narain. 1991. *Global Warming in an Unequal World*. New Delhi: Centre for Science and Environment.

Alaimo, S. 2010. *Bodily Natures: Science, Environment, and the Material Self*. Bloomington: Indiana University Press.

Armiero, M., and L. Sedrez, eds. 2014. *A History of Environmentalism: Local Struggles, Global Histories*. London: Bloomsbury.

Barad, K. 2007. *Meeting the Universe Halfway: Quantum Physics and the Entanglement of Matter*. Durham, NC: Duke University Press.

Bennett, J. 2010. *Vibrant Matter: A Political Ecology of Things*. Durham, NC: Duke University Press.

Braidotti, R. 2013. *The Posthuman*. Cambridge: Polity Press.

Bullard, R. 2000. *Dumping in Dixie: Race, Class, and Environmental Quality*. 3rd ed. Boulder, CO: Westview Press.

Castree, N., et al. 2014. "Changing the Intellectual Climate." *Nature Climate Change* 4:763–68.

Connolly, W. E. 2011. *A World of Becoming*. Durham, NC: Duke University Press.

Corburn, J. 2005. *Street Science: Community Knowledge and Environmental Health Justice*. Cambridge, MA: MIT Press.

Coser, L. 1992. *On Collective Memory: Maurice Halbwachs*. Chicago: University of Chicago Press.

Crist, E. 2013. "On the Poverty of Our Nomenclature." *Environmental Humanities* 3:129–47.

Davis, M. 2001. *Late Victorian Holocausts: El Niño Famines and the Making of the Third World*. London: Verso Books.

Finney, C. 2014. *Black Faces, White Spaces: Reimagining the Relationship of African Americans to the Great Outdoors*. New York: Columbia University Press.

Frawley, J., and I. McCalman, eds. 2014. *Rethinking Invasion Ecologies from the Environmental Humanities*. New York: Routledge.

Golub, A., M. Mahoney, and J. Harlow. 2013. "Sustainability and Intergenerational Equity: Do Past Injustices Matter?" *Sustainability Science* 8 (2): 269–77.

Guha, R. 1989. *The Unquiet Woods: Ecological Change and Peasant Resistance in the Himalaya*. Oxford: Oxford University Press.

Guthman, J., and B. Mansfield. 2013. "The Implications of Environmental Epigenetics: A New Direction for Geographic Inquiry on Health, Space, and Nature-Society Relations." *Progress in Human Geography* 37 (4): 486–504.

Halbwachs, M. 1925. *Les cadres sociaux de la mémoire*, in *Les Travaux de L'Année Sociologique*, Paris: F. Alcan; subsequently published as *Les cadres sociaux de la mémoire*. Paris: Presses Universitaires de France, 1952.

Hartog, F. 2003. *Régimes d'historicité: Présentisme et expériences du temps*. Paris: Seuil.

———. 2013. *Croire en l'histoire*. Paris: Flammarion.

Hornborg, A. 1998. "Towards an Ecological Theory of Unequal Exchange: Articulating World System Theory and Ecological Economics." *Ecological Economics* 25 (1): 127–36.

Hurley, A. 1995. *Environmental Inequalities: Class, Race and Industrial Pollution in Gary, Indiana, 1945–1980*. Chapel Hill, NC: University of North Carolina Press.

Ingold, T., and G. Pálsson, eds. 2010. *Biosocial Becomings: Integrating Social and Biological Anthropology*. Cambridge: Cambridge University Press.

Jamieson, D. 2014. *Reason in a Dark Time: Why the Struggle to Stop Climate Change Failed—and What It Means for Our Future*. Oxford: Oxford University Press.

Jordheim, H. 2014. "Introduction: Multiple Times and the Work of Synchronization." *History and Theory* 53 (4): 498–518.

Koselleck, R. 2000. *Zeitschichten: Studien zur Historik*. Frankfurt am Main: Suhrkamp.

Lynas, M. 2011. *The God Species: Saving the Planet in the Age of Humans*. London: Fourth Estate.

Malm, A., and A. Hornborg. 2014. "A Geology of Mankind? A Critique of the Anthropocene Narrative." *Anthropocene Review* 1 (1): 62–69.

Mansfield, B. 2012. "Race and the New Epigenetic Biopolitics of Environmental Health," *BioSocieties* 7 (4): 352–72.

Martínez-Alier, J. 2002. *The Environmentalism of the Poor: A Study of Ecological Conflicts and Valuation*. Cheltenham: Edward Elgar.

Merchant, C. 1980. *The Death of Nature*. San Francisco, CA: Harper and Row.

———. 1996. *Earthcare: Women and the Environment*. New York: Routledge.

Nixon, R. 2011. *Slow Violence and the Environmentalism of the Poor*. Cambridge, MA: Harvard University Press.

Nora, P. 1989. "Between Memory and History: Les lieux de mémoire." *Representations* 26:7–25.

Oppermann, S. 2013. "Feminist Ecocriticism: A Posthumanist Direction in Ecocritical Trajectory." In *International Perspectives in Feminist Ecocriticism*, ed. Greta Gaard, Simon C. Estok, and Serpil Oppermann, 19–36. London: Routledge.

Pálsson, G., and H. Swanson. 2016. "Down to Earth: Geosocialities and Geopolitics." *Environmental Humanities* 8:149–71.

Pálsson, G., et al. 2013. "Reconceptualizing the 'Anthropos' in the Anthropocene: Integrating the Social Sciences and Humanities in Global Environmental Change Research." *Environmental Science and Policy* 28:4–14.

Robin, L., S. Sörlin, and P. Warde, eds. 2013. *The Future of Nature: Documents of Global Change*. New Haven, CT: Yale University Press.

Røpke, I. 2001. "Ecological Unequal Exchange." Special issue: "Human Ecology in the New Millennium." *Journal of Human Ecology* 10:35–40.

Roy, A. 2009. "Is There Life after Democracy?" Dawn.com, May 7, 2009. http://www.dawn .com/news/475778/is-there-life-after-democracy.

Stoner, A., and A. Melathopoulos. 2015. *Freedom in the Anthropocene: Twentieth-Century Helplessness in the Face of Climate Change*. New York: Palgrave Macmillan.

Swyngedouw, E. 2010. "Apocalypse Forever? Post-Political Populism and the Spectre of Climate Change." *Theory, Culture and Society* 27 (2–3): 213–32.

Wormbs, N. 2013. "Eyes on the Ice: Satellite Remote Sensing and the Narratives of Visualized Data." In *Media and the Politics of Arctic Climate Change: When the Ice Breaks*, ed. M. Christensen et al., 52–69. New York: Palgrave Macmillan.

Zalasiewicz, J., et al. 2018. *The Epochs of Nature: Georges-Louis Leclerc, le Comte de Buffon*. Translated, edited, and compiled by J. Zalasiewicz, A.-S. Milon, and M. Zalasiewicz, with an introduction by J. Zalasiewicz, S. Sörlin, L. Robin, and J. Grinevald. Chicago: University of Chicago Press.

Zylinska, J. 2014. *Minimal Ethics for the Anthropocene*. Ann Arbor: Open University Press.

Objects from Anna Schwartz's Cabinet of Curiosities

Judit Hersko

Let me introduce you to Anna Schwartz, seen here in a silicone portrait from the Stefansson archives in Dartmouth (plate 12). She is an unknown explorer, the only woman to make it onto a US Antarctic expedition before the 1960s. Strangely, Anna Schwartz's emergence from obscurity is not due to this stunning accomplishment. Instead, it is her obsession with two tiny pelagic snails, the sea butterfly (*Limacina helicina*) and its predator, the sea angel (*Clione limacina*), that has made her into a recent emblem of the Anthropocene. The sea butterfly, a shelled planktonic snail, functions as the canary in the coal mine when it comes to ocean acidification, one of the most insidious aspects of anthropogenic climate change.[1]

Schwartz was born in 1920 in Budapest, Hungary. At age twelve, she received a box camera that sealed her fate. Her first photograph was published in a newspaper, and by age fifteen she had developed an obsession with phenomena of light, shadow, and transparency. Schwartz was also an avid naturalist. She admired German-born Maria Sibylla Merian, who was the first person to observe and document the metamorphosis of butterflies. In 1699 Merian voyaged from Amsterdam to Surinam in South America to search for exotic caterpillars, the only woman known to travel for scientific reasons during that era.

It was one of her naturalist friends, a student of oceanography in London, who told Schwartz about the cruise of the HMS *Challenger* (1872–76), the first global Marine expedition organized and funded by the British Royal Society with exclusively scientific purpose. Its mission was to examine the deep-sea floor and answer comprehensive questions about the ocean environment. Based on her friend's account, Schwartz determined that she had to see the *Challenger* manuscripts first-hand at the Natural History Museum in London. She arrived there in 1937, and spent several months in the company of young people she had been introduced to by her friend.

One of her new acquaintances had an aunt, Beryl White, who spent much time in India with her family and created photographic albums that eschewed the customary picturesque distancing present in most representations of British India. She combined watercolor painting and photography in ways that intrigued Schwartz. White's work also introduced her to the tradition of album making that thrived in nineteenth-century England, and Schwartz realized that women of the Victorian era had effectively invented the photo-collage later adopted by avant-garde artists. These Victorian albums were self-representations made for an audience of friends and family. Photographic clippings of relatives, acquaintances, and strangers were placed in painted settings including drawing rooms, the sites of upper-class amusement such as dinners and balls, croquet and lawn tennis, boating and riding, and the circus or theater. Images of people were also collaged onto painted objects such as playing cards, clocks, bags, umbrellas, fans, china, and jewelry, and in some cases they were strewn around or juggled by jesters or incorporated into patterns. In some cases images (often of children) were miniaturized and placed in fairytale environments. Yet other collages depicted human faces on animal bodies that varied from monkeys to ducks and spiders. This latter custom was in response to the recent publication of Darwin's theory and the widespread caricature spawned by it.

Schwartz's favorite collage artist, Kate Edith Gough, was the daughter of a wealthy industrialist. In her sophisticated and humorous collages Gough often combined photographs with watercolor backgrounds in order to place her subjects in fantastic or dramatic environments. Her work inspired Schwartz to experiment with surreal juxtapositions and the placement of people in settings they would not normally inhabit. After returning home from London, she created a series of collages including one entitled "With Scott at the South Pole." She placed a photograph of her twelve-year-old self, wearing summer clothes, in the right corner of that famous image capturing Robert Falcon Scott and his companions as they reached the South Pole and discovered the Norwegian flag planted by Amundsen.

It was in London while poring over the pages of the *Challenger* report that Schwartz first came across the sea angel and the sea butterfly (Pelseneer 1888). The *Limacina helicina* or sea butterfly (here seen embedded in her silicone portrait) is abundant in the cold oceans at high latitudes together with its predator, the *Clione limacina* or sea angel. Both species use wing-like flaps to propel them-selves through water, which is why they are called pteropods, a Greek term for "wing-foot." The sea butterfly belongs to clade Thecosomata (shelled pteropod) while the sea angel is a Gymnosomata, meaning it has no shell. Both organisms are transparent, but while the sea angel can grow to three inches, the largest sea butterfly is under a quarter inch. The fairy-like sea angel feeds almost exclusively on the sea butterfly using highly specialized tentacles that emerge dramatically from its head to pry open the shell of its prey. Although Schwartz only had draw-ings and descriptions to go by she was captivated by the symbiosis of these organ-isms as well as their fanciful winged form and transparency. She wanted to be the first one to photograph the light coming through their bodies and shells, and she was prepared to go to great length to achieve this.

In 1938 Schwartz crossed the ocean to visit her distant cousin Evelyn Schwartz, who was born and raised in New York. At this time Evelyn was involved in a bur-geoning relationship with Vilhjálmur Stefansson, the famous and controversial polar explorer. As a young man, Stefansson had spent extended periods in the far north. Later he made a living by lecturing on the "friendly Arctic" and its poten-tial for human expansion. Stefansson served as the president of the Explorers Club and Schwartz was counting on his help to achieve her quest.

In New York Schwartz joined Evelyn and Stefansson on their regular outings to Greenwich Village where the former explorer regaled the two women with tales of his adventures, claiming that the Arctic is the future territory of human prosperity. In 1921 he had handpicked and sent a team of four young men, three Americans and a Canadian, to create a permanent settlement on Wrangel Island in order to prove his theory and claim the land for Canada. They hired an Inupiat woman, Ada Blackjack, as their cook and seamstress, and she became the sole survivor of the expedition. Stefansson did not emphasize the general failure of this experiment but he touted Blackjack's ability to develop trapping and hunting skills to survive as an example of the friendly Arctic.

While Schwartz was skeptical about Stefansson's claims she secretly hoped that he could help her reach the far north and find the sea angel and the sea but-terfly. For this reason she was intent on learning as much as she could about the region. Stefansson had gifted signed copies of his books to Evelyn, and Schwartz, who had much time on her hand, began reading them. Although she did not enjoy his impersonal writing style, Schwartz grew interested in the Inuit team members of his expeditions and the fact that he lived with them in order to more accurately study their lifestyle. She heard from Evelyn about the blue-eyed Inuk, Alex Ste-fansson, who was rumored to be Stefansson's son with an Inuit woman. However, her cousin confided that Stefansson, whom she called Stef, refused to speak about the boy or about anything else regarding his private life in the field.

* * *

Schwartz immersed herself in research about the Arctic. She learned about the places that featured prominently in Stefansson's adventures such as Herschel Island. The first foreigners arrived there in search of bowhead whales in 1889 and the island became the whaling center of the western arctic from 1894 to 1905. Whale oil was a prized commodity, but more than anything the boom in the whaling industry at that particular moment was driven by the fashion of women's corsets that required baleen for structural support. This keratinous substance, hanging from the upper jaws of baleen whales, functions as the filter for their food intake, which, incidentally, includes the sea angel and the sea butterfly among other planktonic organisms.

Schwartz found it ironic that the constrictive fashion of Victorian women, often featured in her beloved collages from the Victorian era, was the driving force behind the whale trade on Herschel Island, while, according to most accounts, the lifestyle on the island itself was anything but restrained. Foreigners tended to arrive without women and led a raucous existence. They hired Inuit "seamstresses" who performed more than one role in their lives. Some of these men did settle in the Arctic with their Inuit families as, for example, the Norwegian Storker Storkerson, a participant in three of Stefansson's expeditions.

When her cousin showed her a group photograph with Storkerson in the center, Schwartz grew interested in the woman and child on the left side of the image. She knew by now that in 1908 Stefansson had hired a "seamstress," Pannigabluk, and that she was the alleged mother of his son Alex. However, when she made her famous collage superimposing Pannigabluk and Alex onto a corset made of baleen, Schwartz did not know that later researchers would find records indicating that the wife and son of Stefansson, Pannigabluk and Alex, were baptized on Herschel Island in 1915 (Stefansson and Pálsson 2001, 14–15). Nor did she know that Stefansson lived with Pannigabluk on and off for nine years and with Alex for eight, and that Pannigabluk was not just Stefansson's main ethnographic informant but that she conducted many of the interviews and collected much of the material described in his work. She also saved his life and nursed him back to health on several occasions, while performing most of the physical labor required for their existence including hunting and building their shelter, as he sat and wrote in the field. Nor could Schwartz know at this time that Pannigabluk raised Alex alone and never lived with a man again after Stefansson left in 1918, or that

Stefansson, who at age sixty-two had remained a staunch bachelor, would not marry Evelyn Schwartz before 1941, a year after Pannigabluk's death.

* * *

As the president of the Explorers Club Stefansson often chose to preside over club meetings at Romany Marie's restaurant in Greenwich Village. Romany Marie, a colorful character of Romanian Gipsy descent, enjoyed the company of artists, intellectuals, and bohemians. She ran an ordinary diner that hosted extraordinary people. Evelyn began to frequent Romany Marie's while she was still in high school. Much of her life was shaped by the relationships she forged there, including the one with Stefansson. It was here, in the early summer of 1939, at a dinner party loosely associated with the Explorers Club, that Schwartz met Ruth Hampton, a representative of the Department of the Interior on the Executive Committee of the US Antarctic Service designated by President Roosevelt. As they were sitting next to each other their conversation drifted to photography as well as to Schwartz's mission to document the sea angel and the sea butterfly. It was a perfectly pleasant evening but Schwartz did not dwell on it, until a few weeks later when an excited Evelyn knocked on her door. Apparently Ruth Hampton had contacted her through Stef and asked if Schwartz could type. Hampton's department was in charge of hiring volunteers for Admiral Byrd's US Antarctic service expedition. They had a photographer lined up who was a decent typist but he had fallen ill. The departure for the expedition was fast approaching and they were in a pinch for someone to record scientific data. Present in the right place at the right time, and possessing some of the right skills, Schwartz got the job. Although she had originally planned to travel to the Arctic, she knew that this was her great opportunity as the sea angel and the sea butterfly were equally abundant in the southern oceans. She had to disguise herself as a man (no women were allowed on such expeditions), but her worry about this was overshadowed by excitement. Another of her heroes, Jeanne Baret, the first woman to circumnavigate the globe back in 1776, set sail disguised as a male valet and botanical assistant to Philibert Commerson, physician and royal naturalist on Louis Antoine de Bougainville's voyage around the earth.

Jeanne Baret made it all the way to Tahiti before she was revealed to be a woman. Schwartz too reached Antarctica without incident, but soon her disguise as a man proved to be awkward and unsustainable. Her sojourn had to be cut short, and when the ships set sail to return to the US on March 20, 1940, she was onboard. During her brief stay, Schwartz made heroic attempts to capture images of the planktonic snails. While her male comrades were busy preparing to map the vast "heroic" landscape with the aid of airplanes and the monstrous snow cruiser, Schwartz sat huddled at the microscope with the sea angel and the sea butterfly. Despite the fact that in 1934 Frits Zernike had invented the phase-contrast microscope, allowing for the study of colorless and transparent biological materials, she was not able to capture them on film as such equipment was not yet available in Antarctica. Instead, she took to making "invisible" embroideries and drawings with transparent materials that cast "photographic" shadows on the wall when lit from a focused light source.

For now we leave Schwartz in Scott's Terra Nova hut at Cape Evans where she shared the space of former explorers and worked in Herbert Ponting's dark room. Fast-forward sixty-five years to San Diego, California, where Schwartz's daughter, an artist, decided to finish what her mother had begun. She was originally trained in figurative sculpture but, having inherited her mother's preoccupation with light, she later turned to light projections as well as transformations of matter in order to speak about experiences of time, loss, and memory. She appropriated her mother's collages and turned them into mirrors that both reflect and transmit the filmic projections of early polar expeditions, thereby fragmenting the masculine fantasies of heroism and objectivity. She teamed up with scientists studying the *Limacina helicina* and the *Clione limacina* in order to complete her mother's quest. One of the researchers at the forefront of this work is Dr. Victoria Fabry, who examines the effects of ocean acidification on calcifying organisms such as pteropods. Dr. Fabry enticed Schwartz's daughter with stunning images; the dreams of Schwartz come true. Most exciting are the representations of the sea butterfly, including motion pictures taken through the microscope.

Schwartz's daughter read an article Dr. Fabry coauthored in the 2005 September issue of *Nature* titled "Anthropogenic Ocean Acidification Over the Twenty First Century and Its Impacts on Calcifying Organisms" (Orr et al. 2005). This article describes how the increased level of carbon dioxide (CO_2) produced by human activity, primarily the burning of fossil fuels in the manufacturing of products and the driving of vehicles, is causing changes in ocean chemistry. Carbon dioxide does not stay where it is produced but spreads in the atmosphere equally across the globe and is absorbed by the oceans. As a result, chemical reactions occur that reduce seawater pH (hence the term "ocean acidification") as well as the saturation states of biologically important calcium carbonate minerals. These minerals, including aragonite, are the building blocks that organisms such as the sea butterfly use to build shells. Researchers found that if carbon dioxide emissions continue at current rates, the seawater at high latitudes will become corrosive to shells within decades rather than centuries as was previously thought.

When scientists exposed live planktonic snails to predicted levels of pH and calcium carbonate mineral saturation during a two-day shipboard experiment, they observed that the aragonite shells showed notable dissolution. As the CO_2 in the oceans lowers the concentration of aragonite, the sea butterfly, as well as corals and all manners of creatures with shells, are in danger of simply disappearing, with potentially catastrophic consequences for the entire marine food chain. These effects, as many other symptoms of climate change, are occurring most rapidly at the two poles, where the chemistry and temperature of seawater speed up the acidification process.

While both poles beam back alarming data to the center in regards to the state of our planet, their local history is very different. In Antarctica the scientists and their support staff are the only human inhabitants. Their presence on the continent is very recent (slightly more than 100 years) and they do not depend on the environment for their survival but bring their supplies with them. In the Arctic, the Inupiat people, the descendants of Pannigabluk and Stefansson's other team members, come from a long lineage of indigenous inhabitants who consider "the Arctic Ocean as their garden and depend on it for nutritional, cultural and spiritual sustenance." In a recent gathering, Inupiat Elders from Unanakleet to Greenland issued a warning against disturbing the oceans "where our food chain begins" and where it needs to be "protected at all costs" (Banerjee 2015, 27).

The Chukchi Sea is one of the richest and most complex marine habitats on Earth. It serves as a migration corridor for the bowhead whales whose baleen was so sought after on Herschel Island in the age of whale oil and Victorian corsets. Bowheads were decimated, close to the point of extinction, during that era but have since recovered somewhat. Nevertheless, they are on the endangered species list and only "Native Alaskans and Canadian Inuit are allowed a limited subsistence hunt for bowhead whales from stable populations" (World Wildlife Federation n.d.). It turns out that these Arctic whales are the longest-lived wild mammals on the planet. Based on stone harpoon tips retrieved from their blubber, as well as on eye tissue analysis, scientists have determined that their lifespan can be over 100 years (and sometime even over 200). Bowhead whales are currently circling the Arctic carrying the weapons of whalers who arrived on Herschel Island in 1889. Some of them may have leapt out of water to delight Stefansson and his family when Pannigabluk and Alex were baptized on the island in 1915. These giant, ancient mammals share the Arctic waters with all manner of creatures, but they depend for their survival on the short-lived, "tiny subsea creatures that make up the food web but elude our eyes" such as the sea angel and the sea butterfly (Banerjee 2015, 27).

NOTES

1   Anna Schwartz is a fictional character who I have inserted into real historical events with real historical people. The science presented is factual and derives from my collaborator, Dr. Victoria Fabry, who is at the forefront of research on ocean acidification and its effects on calcifying organisms.

BIBLIOGRAPHY

Arctic Monitoring and Assessment Programme (AMAP). 2013. "Arctic Ocean Acidification Assessment: Summary for Policy-makers." Updated May 23, 2013. http://www.amap .no/documents/doc/amap-arctic-ocean-acidification-assessment-summary-for-policy -makers/808 (accessed October 11, 2015).

Banerjee, S., ed. 2012. *Arctic Voices: Resistance at the Tipping Point*. New York: Seven Stories Press.

———. 2015. "In the Warming Arctic Seas." *World Policy Journal* 32 (2): 18–27. doi:10.1177/0740277515591539.

Bloom, L. E. 1993. *Gender on Ice: American Ideologies of Polar Expeditions*. Minneapolis: University of Minnesota Press.

———. 2015. "Judit Hersko's Polar Art: Anthropogenic Climate Change in Antarctic Ocean-scapes." *UCLA Center for the Study of Women Newsletter* (Summer). http://escholarship.org/ uc/item/7jx8m9gb.

Bravo, M., and S. Sörlin, eds. 2002. *Narrating the Arctic: A Cultural History of Nordic Scientific Practices*. Canton, MA: Science History Publications.

Comeau, S., G. Gorsky, R. Jeffree, J.-L. Teyssié, and J.-P. Gattuso. 2009. "Impact of Ocean Acidification on a Key Arctic Pelagic Mollusc (*Limacina helicina*)." *Biogeosciences* 6:1877–82. doi:10.5194/bg-6-1877-2009.

Comeau, S., R. Jeffree, J.-L. Teyssie, and J.-P. Gattuso. 2010. "Response of the Arctic Pteropod *Limacina helicina* to Projected Future Environmental Conditions." *PLoS ONE* 5 (6): e11362. doi:10.1371/journal.pone.0011362.

Dohmen, R. 2012. "Memsahibs and the 'Sunny East': Representations of British India by Millicent Douglas Pilkington and Beryl White." *Victorian Literature and Culture* 40 (1): 153–77.

Hersko, J. 2009. "'Translating' and 'Retranslating' Data: Tracing the Steps in Projects that Address Climate Change and Antarctic Science." *Digital Arts and Culture 2009*. Digital Arts and Culture 2009, University of California Irvine. http://escholarship.org/uc/item/ 40z2b75n (accessed October 11, 2015).

———. 2012. "Pages from the Book of the Unknown Explorer." In *Far Fields: Digital Culture, Climate Change, and the Poles*, ed. Andrea Polli and Jane Marsching, 61–75. Bristol: Intellect Books.

Hunt, B., J. Strugnell, N. Bednarsek, K. Linse, R. J. Nelson, E. Pakhomov, B. Seibel, D. Steinke, and L. Würzberg. 2010. "Poles Apart: The "Bipolar" Pteropod Species *Limacina helicina* Is Genetically Distinct between the Arctic and Antarctic Oceans." *PLoS One* 5 (3): e9835. doi:10.1371/journal.pone.0009835.

Jørgensen, D., and S. Sörlin, eds. 2013. *Northscapes: History, Technology, and the Making of Northern Environments.* Vancouver: UBC Press.

King's College Online Exhibitions Archives. "A Daughter of the Empire: Beryl White in India, 1901–03." http://www.kingscollections.org/exhibitions/archives/a-daughter-of-the-empire/ (accessed October 11, 2015).

Kolbert, E. 2011. "The Acid Sea." *National Geographic* 219 (4): 100–103.

———. 2012. "The Darkening Sea: What Carbon Emissions Are Doing to the Ocean." In *The Global Warming Reader: A Century of Writing about Climate Change*, ed. Bill McKibben, 377–98. New York: Penguin.

Leonardo Electronic Almanac. "Anna's Cabinet of Curiosities" by Judit Hersko. http://www.facebook.com/media/set/fbx/?set=a.10150147769611253.346156.209156896252 (accessed October 11, 2015).

Niven, J. 2003. *Ada Blackjack: A True Story of Survival in the Arctic.* New York: Hyperion.

Orr, J. C., V. J. Fabry, O. Aumont, L. Bopp, S. C. Doney, R. A. Feely, A. Gnanadesikan, et al. 2005. "Anthropogenic Ocean Acidification over the Twenty-First Century and Its Impact on Calcifying Organisms." *Nature* 437 (7059): 681–86.

Pálsson, G. 2003. *Travelling Passions: The Hidden Life of Vilhjalmur Stefansson.* Hanover, NH: UPNE.

Pelseneer, P. 1888. *Report on the anatomy of the deep-sea Mollusca collected by HMS Challenger in the years 1873–76.* Gt. Brit. Challenger office. http://www.19thcenturyscience.org/HMSC/HMSC-Reports/Zool-66/htm/doc.html (accessed October 11, 2015).

Riebesell, U., J.-P. Gattuso, T. F. Thingstad, and J. J. Middelburg. 2013. "Arctic Ocean Acidification: Pelagic Ecosystem and Biogeochemical Responses during a Mesocosm Study." *Biogeosciences* 10:5619–26.

Siegel, E. 2009. *Playing with Pictures: The Art of Victorian Photocollage.* Chicago: The Art Institute of Chicago/Yale University Press.

Stefansson, V. 1921. *The Friendly Arctic: The story of Five Years in Polar Regions.* New York: Macmillan.

———. 1922. *My Life with the Eskimo.* New York: Macmillan.

Stefansson, V., and G. Pálsson. 2001. *Writing on Ice: The Ethnographic Notebooks of Vilhjalmur Stefansson.* Hanover, NH: UPNE.

Stefansson Nef, E. 2002. *Finding My Way: The Autobiography of an Optimist.* Washington, DC: Francis Press.

World Wildlife Federation. N.D. "Bowhead Whales." http://wwf.panda.org/what_we_do/endangered_species/cetaceans/about/right_whales/bowhead_whale/ (accessed October 15, 2015).

Technofossil

Jared Farmer

As an exercise in environmental speculation—and amateur art—I created a future fossil of a BlackBerry Curve 8300, a gadget mass introduced in 2007, since discontinued. I took a piece of Utah mudstone from my late father's rock collection and adhered onto it a layer of polymer clay, a petroleum byproduct, into which I made an imprint using a broken BlackBerry purchased on eBay for small change. Then I baked it, and painted it (plate 13). Given that the progenitor of my device was Research in Motion (RIM), and given that the genus name for the blackberry plant is *Rubus*, I named my specimen *Rimus rubus curvus* (2014).

Without a field guide, few children or teens could identify my fossil, though for a pop culture nano-epoch its design was globally iconic. The BlackBerry population that succeeded the Curve, the Bold, now faces its own discontinuation. In 2014 professional celebrity Kim Kardashian revealed her private conservation strategy: She buys Bolds on eBay and hoards them. "I have anxiety that I will run out," she said. "I'm afraid it'll go extinct."

On the surface, my art piece satirizes the relationship—verbal and material—between consumer capitalism and biological extinction. Advertisers have brazenly coopted the language of ecology and evolution to naturalize planned obsolescence, extending the 1960s idea of "product life cycle." Products must evolve—or die. Technology journalists create listicles of endangered gadgets; they warn us that even adaptable ones can go extinct if they live within innovation-starved ecosystems; they refer to outmoded devices as dinosaurs or living fossils. Consumers have been trained to expect, even celebrate, mass extinctions (e.g., Zune, Palm, Nook) as necessary and inevitable functions of the market. Compared to conservation red lists, the anxiety provoked by technology doom lists is ameliorated by the pleasurable anticipation of the next "generation" release, and the agreeable pre-nostalgia for the future vintage. Defunct gadgets—especially the media players of our youth—are nicer to consider than destroyed species.

The genius of the iPhone (introduced in 2007, the same birth year as my Curve) was to make personal computing more personal—bodily, intimate—yet more disposable. By 2014 the number of mobile Internet devices surpassed the number of people on Earth. By the mid-twenty-first century, effectively every adult will have the equivalent of a first-generation supercomputer on (in?) their person at all times. First adopters will have gone through scores, even hundreds, of gadgets before the last unconnected human gets equipped. In a consumerist corollary to Moore's law, savvy users expect to upgrade no less than every two years—fast fashion for hardware. "Future shock" has been routinized.

Constant upgrading demands coordinate degradation: the newest new thing is simply the latest future trash. According to EPA estimates, only 8 percent of the 129 million cellphones discarded by Americans in 2009 were recycled. Most were insourced to local landfills, where the detritus of the information age accretes and leaks. Compared to stone tablets, or papyrus scrolls, or leather-bound paper books, digital texts are perilously ephemeral. And yet all the containers for all the circuits that briefly held the underlying binary code can be astonishingly persistent.

Dead media are in fact undead. The ever-shorter lifespans of consumer electronics belie their ever-longer pathways. The message on the back of each iPhone—"Designed by Apple in California. Assembled in China"—could include an addendum, "Extracted from Africa." Despite the Basel Convention, the supply chain sometimes goes full circle: salvagers in toxic workscapes in countries such as Ghana and Nigeria quarry circuit boards by hand. Theirs is a grey economy, not a green one. Disassembly is haphazard and hazardous because electronic gadgets were not designed to be recycled. Silicon Valley has yet to disrupt the economy of wastefulness. The tech industry has no incentive to acknowledge that its cumulative externalities have grown bigger with miniaturization. When Steve Jobs gave his now-canonized 2005 commencement speech at Stanford, he ignored the airplane dragging a banner through the cloudless California sky: STEVE, DON'T BE A MINIPLAYER—RECYCLE ALL E-WASTE.

If my fossil signifies extinction and e-waste, it also bespeaks a cultural moment when humanists and artists have synchronously turned their attention to temporalities like "deep time," "deep history," "big history," "long-termism," and "the long now." Predictably, one critical anthology calls it the "geologic turn." Following the trend, universities have convened symposia on geoaesthetics, geoculture, and planetarity. In the art world, a representative work is Julian Charrière's Berlin exhibit *On the Sidewalk, I Have Forgotten the Dinosauria* (2013), which horizontally displayed an eighty-meter vertical core sample. Gallery visitors walked through time—fossils zones to paving stones—as they witnessed layers normally hidden below their feet. By referencing climate proxies such as ice cores, Charrière's novel work of found art suggested a parallel between drilling deep for data and looking deep for wisdom.

After a period of cultural quiescence, geology—the leading realm of inquiry in the late eighteenth and early nineteenth centuries—has regained philosophical, literary, and artistic significance. The new geotheory tracks with "Anthropocene," a keyword that erupted in 2000, forcing its way into the *OED* in 2014. The discourse about the Anthropocene can be imagined as two overlapping spheres of unequal size: (1) a limited geological conjecture about the lithosphere—more specifically, the stratigraphic record—as it will exist millions of years from now; and (2) a limitless rhetorical description of human influence on the complete Earth System (lithosphere, hydrosphere, atmosphere, biosphere) in the current moment and the recent past.

The International Commission on Stratigraphy has definitional authority over the first concept—the proposed end of the Holocene ("Recent Epoch," literally *Wholly New*) and the beginning of the Human Epoch. Since 2009, a working group led by Jan Zalasiewicz has been gathering pro-and-con (overwhelmingly pro-) evidence for a formal revision of the geologic time scale, otherwise known as the International Chronostratigraphic Chart. Zalasiewicz, a paleobiologist, has speculated that a future "Urban Stratum" of the earth's crust will have new types of *ichnofossils*: lithified records of biological activity. For example, subway tunnels—worm tracks of mammoth size—might become sedimentary molds for locomotion traces (a subclass of ichnofossils called *repichnia*).

However impressive our bioturbations may be and become, they may not be widespread enough to effect a legible layer in the future global rock record. The Anthropocene Working Group is looking for the clearest "signature" onto which they, and also far-future (alien?) forensic scientists, can place a stratigraphic "golden spike." (Under current convention, each situated spike—actually bronze, not gold—is officially called a Global Boundary Stratotype Section and Point.) For example, the fossilized bones of trillions of broiler chickens will probably compose a distinctive fossil layer, a proxy for the "Great Acceleration" of postwar global change. Another prime candidate for geologic marking is the radionuclide signature left by atomic fallout in the 1950s. Across the globe, fallout from above-ground testing deposited isotopes of plutonium—an exceedingly rare element—that will likely be detectable and datable far into the future.

In the here and now, the signal of radiation has a signification problem: it is intangible, invisible. The preemptive search for the next golden spike is more about communicating science than science per se. Ideally, the Anthropocene's signature would be legible to the public. Various interest groups—for and against geogovernance, for and against geoengineering—have a stake in the PR of geology. In this political debate surrounding the stratigraphic deliberation, high-level radiation is background noise. The Cold War—and the immediate threat of flash devastation and impact extinction—simply (thankfully?) doesn't resonate as it once did. Climate change no longer means nuclear winter. At least for the moment, humble carbon, oxygen, and hydrogen induce more catastrophic thinking than transuranic plutonium. A Cold War geomarker can't fully capture our current anxieties about the post-Wholly New—an Earth System in which creeping change can produce state shifts, a weird world, no longer improbable, that melts, subsides, desiccates, and inundates.

In 2014, in the first issue of the first volume of the *Anthropocene Review* (one of three new journals), Zalasiewicz and his controversial working group outlined a category of speculation with greater popular resonance: "technofossils." Zalasiewicz et al. suggested that various postwar consumer artifacts could become georeferences because of rapid and wide proliferation. Speculative geologists can choose from an embarrassment of drossy icons: toilets, toothbrushes, beverage cans, paperclips, ballpoint pens. The most important of these proto-fossils may be the lowliest—throwaway plastic containers. High-tech devices such as smartphones also have fossil potential—for rare materials (like uncombined aluminum), for unknown materials (like uncombined vanadium), and for novel materials (like polypropylene). Roughly speaking, a smartphone is 40 percent metal, 40 percent plastic, 20 percent ceramic. Each is a treasure box of rare earth materials.

Will the Great Acceleration be remembered—and recorded in stone—for its low-tech trash or its e-waste? How will our geologic selfies expose themselves? We can only speculate. At the onset of the "paperless revolution" in the 1990s, librarians and archivists imagined a future "Digital Dark Age" of unreadable data and unusable devices. Turning that notion on its head, archaeologists and geologists now predict a future golden age of high-tech middens. Even as lightweight plastic bags are being phased out in many countries and provinces, mobile-connected devices—the first iteration of wearable computers—are still in their expansion phase. All that mobility and wirelessness requires mounds of infra-structure euphemized as "the cloud."

The main communicative mode of the Anthropocene—and late-stage glo-balization, arguably the same thing—is speculation. According to the neoliberal consensus, the future of the planet, just like the business world, depends on our ability to model, predict, invest, and hedge. Even critics of "market fundamen-talism" embrace data-driven futurology. For example, in *The Collapse of Western Civilization: A View from the Future* (2014), historians of science Naomi Oreskes and Erik M. Conway use the cli-fi genre to prognosticate the consequences of ignoring predictions.

From the gadget's-eye view, the prospect is different—neither dystopian nor ecotopian. To describe the long legacies, widespread impacts, and many afterlives of digital data and their material containers, theorists-cum-garbologists have coined neologisms such as "residual media," "fossil media," and "zombie media." For these scholars, the material of the media is the message. Borrowing from dis-ciplines accustomed to long-term and deep-time thinking with objects and sites, "new materialists" sometimes call their work "media archaeology" (after Sieg-fried Zielinski's "Medienarchäologie") or "media geology" (Jussi Parikka's phrase [2015]). Charting a course toward a new philosophy—"infrastructuralism"—John Durham Peters (2015) ponders the condition in which media are elemental and environments are media. His basic point is not arcane. Familiar brand names such as MacBook Air and Amazon Fire suggest how natural it seems to (re)equate media with planetary infrastructure.

In the art world, a parallel movement is happening: a number of contempo-rary artists terraform objects that play with anthro-geology and geohumanities. If not quite a meme, the future fossil has been trending. My amateur art has pro-fessional company.

Consider *Télofossiles II* (2015) by Grégory Chatonsky and Dominique Sirois, exhibited in Taipei and later Beijing. The artists collected ordinary objects, scanned them with Microsoft Kinect sensors, created molds with 3D printers, and filled the molds with earthen material. Chatonsky explained that they wanted viewers to consider species-level mortality. By returning a hard drive, or a shoe, to its essential "minerality," the artists make mass petrification personal. Any arti-fact can become a modern memento mori through this artistic and intellectual process. The duo followed with *Extinct Memories II*, exhibited in Brussels, which imagined a data center excavated after the death of humanity.

Brooklyn-based artist-architect Daniel Arsham, like Sirois and Chatonsky, references the excavated Pompeii—a landscape of the long eighteenth century where the boundaries between the historical, the archaeological, and the geological blurred in a way that now seems oddly contemporary. Arsham's large-scale installation *Welcome to the Future* (2014) consisted of hundreds of used gadgets. He ordered them on eBay, cast them in materials such as volcanic cinder, and placed them in a jackhammered trench in the floor of a gallery. Arsham's color palate runs from ash gray to charcoal gray—an aesthetic reminiscent of Steve Jobs. A man of his time, Arsham makes a sideways critique of capitalism while practicing art-star entrepreneurialism. For his *Future Relic* series, he sells limited editions of everyday objects meticulously lithified to a state of "decayed vintage," not unlike a luxury pair of distressed sneakers. Arsham first received international attention after pop star Pharrell Williams ordered bespoke sedimentary copies of his formative instrument, the 1980s-era Casio MT-500 keyboard/drum kit.

Less famous artists, including Christopher Locke (*Modern Fossils*) and the duo Rebecca Johnson and Jeff Klarin (*Future Fossils*), create and sell faux Stone Age versions of tech products associated with Gen-Xers: Polaroid cameras, Atari joysticks, turntables, and so forth. Such fossils may be less about geological thinking than hipster capitalism: the curation of consumer nostalgia. The hoopla around the 2014 partial excavation of the "mass burial" of hundreds of thousands of unsold Atari cartridges at a landfill in Alamogordo, New Mexico, speaks to this mentality. The Smithsonian Institution accepted a recovered copy of *E.T. The Extra Terrestrial* (one of the biggest commercial failures in video game history) for its permanent collection.

While artists like Arsham use deep historical thinking to further aestheticize, even fetishize, exterior product design, more provocative artists such as Susan Stockwell defamiliarize our devices by revealing their inorganic innards and putting them out of place. *Flood* (2010), Stockwell's site-specific installation at St. Mary's, a fourteenth-century-church-turned-museum in York, England, featured six tons of used computer components that, depending on one's perspective, poured down to the nave from the rafters, or rose from the ground like a media tower of Babel. Less monumentally, Scottish artist-activist Julia Barton's crowdfunded "Littoral Art Project" included a *Future Fossil Collection* (2013) that placed beach litter on the underside of plaster-cast cobbles: demoniacal geodes.

Stockwell and Barton share a sensibility with "circuit benders" and "e-waste artists" who make DIY objects (shown on Pinterest, sold on Etsy) with discarded peripherals, optical disks, and the like. For example, Peter McFarlane's work includes animal forms embedded into circuit boards, a maker-movement simulacrum of fossil-rich sedimentary rocks. Circuit bending and e-waste art have less in common with corporate recycling than hacking. These and other "upcycling" activities compose a recirculating current that, like an eddy, runs alongside but against the main river of consumer capitalism—the one-way flow from mass extraction to mass production to mass consumption to mass elimination. Upcycling can happen at home or at the dump. Since 1990 the company that processes San Francisco's waste stream has sponsored an artist-in-residence program—a practice recently replicated at landfills in Philadelphia, and Portland, Oregon.

The renewed conversation between art and science—one of the notable conditions of the Anthropocene—can be seen in galleries run by scientists themselves. The National Academy of Sciences mounted an exhibit called *Imagining Deep Time* in 2014. The same year, the American Association for the Advancement of Science hung Erik Hagen's *Fossils of the Anthropocene* at its headquarters in Washington, DC. Hagen created his mixed-media "fossils" by embedding everyday objects and plastic waste into canvases painted with a mixture of latex, sand, and marble dust. The result was equal parts appealing and disquieting.

It's unclear if the art of the Anthropocene will produce a unified message. Environmental objects can be troubling; they can be whimsical. In 2015 Dutton Gallery in Manhattan hosted an international group show called *Future Fossils* that explored "ideas of erosion, decay, adaptation in the face of climate change, and the afterlife of objects." The same year, Kelowna, Canada, placed *Fossils from the Future* throughout its downtown arts district with an invitation to tourists to locate and share the mock time capsules with their mobile devices using the hashtag #futurefossils.

The technofossil—both the idea and artistic representations of it—is plastic in its communicative value. What do we see: Geology or biology? Industrial design or consumer waste? My BlackBerry museum piece replicates the binary qualities of digital technology: obsolescent/persistent, ephemeral/permanent, erased/archived, discarded/preserved. An object of ambivalence, the future fossil can inspire bipolar emotional states: wonderment about our Promethean abilities (which allows the possibility of a "good" or even "great" Anthropocene), and dread about our Icarian propensities (which presumes the unavoidability of a bad or worse epoch).

In its amalgam of the human and nonhuman, the future fossil recalls something predigital and even pre-analog: the contents of a Cabinet of Curiosity. In early modern Europe, when aristocratic naturalists filled their *Wunderkammern*, they collected freak forms embedded in rock: unclassifiable *lusus naturae*, or sports of nature. Many such objects in these proto-scientific mini-museums were actually works of artifice. Tellingly, *Wunderkammern* (chambers of wonders) were also called *Kunstkammern* (chambers of art). On their shelves, collectors played with the categories of *artificialia* and *naturalia*. Artisans used corals for elaborate sculpturing and nautilus shells for intricate engraving. They presented fossils as ornaments, talismans, or freak natural objects—or all the above.

No one should assume that technofossils of the deep future—should they occur—will look anything like Paige Smith's *Geode in a Can* (2015) or any other *lusus naturae* by contemporary artists. Future bacteria may evolve (though bioengineering?) to devour our plastics. Over durations of time beyond human reckoning, geologic alchemy and tectonic recycling will transpire. Submerged computer components will, under intense heat and pressure, break down to their molecular parts: ashes to ashes, dust to dust, hydrocarbon to hydrocarbon. Our only immaculate remains will be extraterrestrial, beyond earthly time: the space junk trapped in graveyard orbit, the rovers and Hasselblads abandoned on the moon.

Rather than a predictive object, the future fossil is a physical metaphor, or a rhetorical device in material form. As such, it can help simplify and symbolize anthropogenic effects. Things such as biodiversity loss and global warming are examples of what Timothy Morton (2013) calls "hyperobjects"—effects whose spatial and temporal dimensions are so vast, we can scarcely grasp them despite being caught up in them. You can't put your eyes or hands on climate change, or fossil fuel reserves. A future fossil, by contrast, can be compact, concrete, relatable. It's more like a carbonized handprint than a carbon footprint.

Leaping over the threshold of the Anthropocene, should we hope for or against the unknown, for or against the all-knowable? As prediction and speculation increasingly assume the mantle of prophecy and divination, humanity runs the risk of limiting the imaginary. The determination of the future planet as a permanent exhibition of anthropogenic design may occlude an alternative conception: the current Earth System as an ongoing work of sympoietic art.

BIBLIOGRAPHY

Davis, H. M., and E. Turpin, eds. 2015. *Art in the Anthropocene: Encounters among Aesthetics, Politics, Environments and Epistemologies.* London: Open Humanities Press.

Ellsworth, E. A., and J. Kruse, eds. 2013. *Making the Geologic Now: Responses to Material Conditions of Contemporary Life.* Brooklyn, NY: Punctum Books.

Haraway, D. J. 2016. *Staying with the Trouble: Making Kin in the Chthulucene.* Durham, NC: Duke University Press, 2016.

Morton, T. 2013. *Hyperobjects: Philosophy and Ecology after the End of the World.* Minneapolis: University of Minnesota Press.

Parikka, J. 2015. *A Geology of Media.* Minneapolis: University of Minnesota Press.

Peters, J. D. 2015. *The Marvelous Clouds: Toward a Philosophy of Elemental Media.* Chicago: University of Chicago Press.

Waters, C. N., et al., eds. 2014. *A Stratigraphical Basis for the Anthropocene*. London: The Geo-
    logical Society.
Zalasiewicz, J. 2008. *The Earth after Us: What Legacy Will Humans Leave in the Rocks?* Oxford:
    Oxford University Press.
Zielinski, S. 2006. *Deep Time of the Media: Toward an Archaeology of Hearing and Seeing by Tech-
    nical Means*. Cambridge, MA: MIT Press.

Davies Creek Road

Trisha Carroll and Mandy Martin

MANDY MARTIN (MM): The centerpiece of our painting, *Davies Creek Road* [plates 14a–14b] is the goanna, or monitor lizard, which is one of Trisha Carroll's Wiradjuri totems. Our painting alludes to the waves of Anthropogenic extinction that have exacted a slow violence on our valley in the Central West New South Wales, Australia. Current plans are to build a dam and flood this valley, thus destroying local ecosystems, including that of the goanna and a heritage listed cave system with 6,000-year-old human remains. The flooding will also destroy several neighbors' farms and houses, including Trisha's. Where we live is hotter and drier because of climate change, but building more dams will not make it rain or "drought proof" our region to benefit irrigators and the mining industry.

The goanna (*Varanus varius*) is widely distributed across Australia, often growing to one to two meters. Goannas are frequent summer visitors in our part of the world and love eating our hen eggs as well as small native chicks and eggs and small mammals hidden in tree hollows. Every year we wonder if it will be the last time we see them.

The goanna is part of the dreaming or Jukurrpa of the Wiradjuri people. They are good bush-tucker but because it is Trisha's totem she is forbidden to eat it. There are Wiradjuri scar trees near where we live, carved with goanna forms. These would have marked a boundary of Wiradjuri lands on the Lachlan River. The Wiradjuri name for goanna is Girr-away and there is a Dreaming story about how the goanna got its colors (see fig. 9).

Figure 9. Trisha coordinated a large project painting the pylons of the Cowra Bridge on the Lachlan near where we live. She painted her Wiradjuri totems and significant images including the goanna in the X-ray style. She has painted the goanna in *Davies Creek Road* in this same traditional manner. The large connected dots to the right of the goanna on the pylon represent all the towns located along the Lachlan River, many of which had missions where Aboriginal people were forced to live after violent European colonization of their land and now voluntarily have elected to stay. Trisha's parents and family lived in shacks just near here on the river. Photo: Mandy Martin.

TRISHA CARROLL (TC): The land in the background is very dry. This was made in a time of a lot of drought in this area, ten years ago. When there's drought, the animals just die. And that is what the X-ray (style) is about. There is no flesh left on them. When any animal dies all you see is bones. And this is what happens. The male is on the top of the painting and the female is on the bottom. They were probably wanting to mate and of course had no food or water, so just died.

MM: The painting itself is made in pigments gathered in the area: red is ochre from the end of Davies Creek Road, clay soil with oxides; the deep purple oxide also. Otherwise it is titanium white and black charcoal.

TC: I start with a blank canvas and look at it for a long time. The ideas just come to me, in my brain. I don't think anyone is talking to me—people would think I'm crazy. And then I start to paint and my fingers just cannot stop. And I just don't stop until it's finished. There are some places I've been to before I met Mandy, take a few steps to a place and it's like no you're not allowed in there. And if I take a few more steps the feeling gets stronger. And I've got to leave the place, because it could be dangerous to me. It could be a burial spot, like for my ancestors.

MM: And there is a process involving contact with landscape, sites, and ancestors. There is probably a territory the size of Germany in fact, where Wiradjuri are dominant. Trisha's notion of the dream time is strong. But she is also part of the stolen generation. It is a sad part of Australian colonization.

Deep Time Dreams and Present Nightmares

About an hour's drive from where we live is Bigga Rocks on the tributary of the Lachlan. A large red rainbow serpent snaking over the rock face can also be read as the Lachlan River and is surrounded by other dreamtime figures or Wandjinas, emus, and kangaroo, totems of the Wiradjuri. This site is estimated to be at least 600 years old but likely to have been occupied by indigenous people for millennia.

Our local creeks, Davies Creek and Limestone Creek, flow into the Belubula River, which is part of the Lachlan River catchment in Central Western New South Wales and in turn is connected to the largest water catchment in Australia, the Murray Darling, which eventually after many, many miles runs out to sea in South Australia. We live in the driest state in the driest continent in the world, which almost totally relies on water from the Murray Darling system. During recent record-breaking dry periods, the Murray virtually dried up and ceased to flow in the lower reaches, which is the point where Adelaide, the capital of South Australia, draws its water. There has been a cap for a long time on the Murray Darling Basin for building dams because it already has capacity storage for the natural allocation of rain.

The sad saga of extinction which has occurred since second settlers arrived in this part of the world two hundred years ago has erased much of the population, culture, and language of the Wiradjuri and the ecosystem with which they cohabited. The white tree in our painting, *Davies Creek Road*, is an isolated White Box and only 3 percent of the Grassy White Box Tree community survives at present. Interdependent species like the Superb and Swift Parrots are also on the

Commonwealth Threatened Species list. Other species which were common to our area like the emu, koala, wombat, platypus, glider possums, echidnas, and goanna are rarely seen and many are listed as locally endangered.

The new nightmare is a dam which is proposed to be built in the next five years under a multimillion-dollar initiative of the federal government called the Water Security Regions Program to "drought proof" and secure water supplies. In this instance the Needles Gap Dam, originally scoped thirty years ago, is now one of two sites being proposed. Not one landowner—including irrigators downstream or cattle graziers like us whose land the water will just flood in small part—was consulted. And the Traditional Owners are invisible in the debate.

The new dam scenario would flood this valley, submerging Trisha's house, which is an old shearer's cottage, her veggie garden and hens, her landlord's house and his shearing shed, his son's house and the nearby polo field, and so on. Near that shearing shed is the entrance to a large, heritage limestone cave system which runs from under our property down to the river. This will all be flooded. Yet—we will have water views!

That our government can contemplate building more dams in this rapidly escalating period of climate change boggles the mind. We are seeing extreme variability in our climate and longer droughts, frosts out of season and an over-all drying of the landscape. Our property, which had one the safest rainfall patterns in Australia, now feels marginal. As grass farmers raising cattle, it makes every year increasingly difficult and unpredictable. This time, in 2014, bush fires raged all around us and we helped destroy cattle on the neighbor's property after they were burnt. In 2015, at the same time, we had a snowstorm closing the same roads, sending lizards and snakes which had just come out of winter hibernation hurrying back to cover.

Australia already has more water-storage capacity than we can actually fill; building dams doesn't make it rain more. Our federal government, as you may have heard, are climate-change deniers. Our prime minister believes coal is good for humanity. Our elected government is involved in a spiteful vendetta against scientists and anyone seen as green or from the left side of politics. To placate their unhappy party coalition members of Parliament, there has been a dramatic and retrograde redrafting of the politics surrounding water in the past few years which has massive implications for all our rivers. The dam project and resulting flood look set to go ahead, after the recent weakening of environmental regulations. And this despite its effects on threatened microbat species living in the National Heritage–listed caves within the flood zone. Significant fossiliferous Ordovician limestone deposits also face flooding: these include trilobites and thirty-two newly discovered species of fossil coral. All this and more stands to be lost.

Short-term greed for power and gain looks like it will win again over clear thinking about how Australia is meant to survive—one wonders how we can happily extinguish our last interrelated social systems and ecosystems while thumbing our noses at the obvious signs of rapid climate-induced water shortages.

Anthropocene Cabinets of Curiosity

Objects of Strange Change

Libby Robin

What is the significance of Cabinets of Curiosities in the Anthropocene moment? Museums and artists of the present are returning to cabinets almost like those of the pre-Enlightenment era, as they seek to make sense of the chaotic changes of our present "strange times." *Wunderkammern* juxtapose unlikely things (Impey and MacGregor 1985). They stack and array, they align and contrast. Each object is a counterpoint to other objects, in conversation and contradistinction. Objects in museums have always carried stories across generations and places, drawing out memories of other times. In the twenty-first century, artists too have grabbed the cabinet concept and are using it to invoke the curious anew. While the traditional museum seeks out "authentic" objects in its cabinets, and orders them according to rules, the artist creates curious objects and places them on shelves to disrupt order. Playing with the order and disorder of Cabinets of Curiosity, using the juxtaposition of objects, is an art form. It has now also become a tool to explore the Great Acceleration of change at the apocalyptic turn of the millennium. Cabinets are about the relations between objects: they can be intellectual (the classic natural history cabinet of species), but more often the twenty-first-century cabinets are personal and emotional, a throwback to times before the world was ordered in binomials, to a time when mystery and myth also played a major role in understanding nature and the world at large. A Cabinet speaks of collecting, curating, coordinating and classifying. It offers a way to unpack the curious in strange times. At the end of the Holocene epoch, it has become a lens on the world in this rupture of time.

Ole Worm's *Wunderkammer*

In the city of Aarhus in 1655, the collector and antiquarian Ole Worm created a "museum," a room of wonders, curious objects that challenged human understanding of the systems of the world, of the way the world worked. Each item was there because it was curious, not merely weird as we might say today, but rather "not yet fully understood," or literally *stimulating of curiosity*. His chamber of wonders, engraved in a famous illustration, reveals various relations between his chosen objects. Juxtapositions reflect relatedness. There was a wall of antlers, classic hunting trophies, with horns and leg bones beneath. But the long horn of the Narwhal was over on a shelf with shells and creatures of the sea. Polar bears and sea creatures hung from the roof (the north is at the top, perhaps?) and the human tools and clothing also belonged to people of the north, the trade-links between Denmark and Greenland, which was the "new world" of Worm's day. Each item challenged the categories of what ought to be in the *natural* world—it was both of nature and uncanny. These objects together had an aura, a magic that made them entertaining. The arrangements of Worm's objects are still surprising and unsettling even today.

Antiquarians like Worm have been patronized, seen as merely parochial: their systems seem unsystematic, even promiscuous. Philosopher Friedrich Nietzsche wrote scathingly in 1874 about antiquarians who collected "all that is small and limited, mouldy and obsolete," and then used their collections to define themselves and their towns (quoted in Griffiths 1996, 2). Yet a collection is a clue about organization strategies that are vernacular. They are based on personal taste, not on universal principles, and they speak back to a scientific order that claims to universalize the world. Curiosity is personal and is essential to scientific endeavor, even as science strives to transcend the personal and the "merely local." Worm's collection reflected what *mattered* to him as a collector at his particular time. His *Wunderkammer* was stacked with "wonder-full" objects, exemplars of what *surprised* him as a collector. His cabinet was no exhaustive audit of what was already known: it was rather a place of emotional investment, lovingly tended by its creator. Worm did not just array objects to explore an intellectual order, but, rather, he chose objects because they were *curious*, in every sense of the word.

Cabinets may offer keys to the nature of curiosity itself, to new ways of thinking about the world. Worm's museum offered an order of nature a century before another Nordic thinker, Carl von Linné, developed the classification of plants and animals that became the basis for Western scientific nomenclature, a system for "everywhere." A Cabinet of Curiosity is intensely local, focused and is not designed to provide systemic structure, yet it can be revealing of the thought patterns of its curator. The *Musei Wormiani Historia* was a project of its own time, yet it framed many of the questions that Linnaeus later approached in a different way. From our present day perspective, Linnaeus provided such a powerful order that we still use his system. We still see the world Linnaeus's way—and find Worm's juxtapositions odd; they are weird rather than curious. Yet as the world itself undergoes rapid change, at a time when the order of the Holocene is challenged by the emergence of a new geological epoch, we are finding new ways to wonder at curiosities.

Cabinets as High Art

At a time when objects no longer come from strange places because the world is well known, there is a new turn to *Wunderkammern*. Rather than amateur or antiquarian in style, Cabinets of Curiosity are emerging in high art, as tools for exploring our strange and rapidly changing times, for expressing the vortex of Great Acceleration. Artists build cabinets to array their created objects in ways that provoke thinking and rethinking: they celebrate neither the natural nor the unnatural, but play with curiosity itself.

OUT OF THE DARKNESS

I traveled in September 2015 to Ole Worm's town of Aarhus in Denmark and encountered *Out of the Darkness*, a conceptual exhibition of contemporary art at the ARoS Art Museum.[1] *Out of the Darkness* is curated by Erlend G. Høyersten—or perhaps I should say "directed," as it is a performative and conceptual exhibition that messes with perceptions and space and time, as well as light and shadow. The visitor herself becomes the player on the strange stages of Høyersten's labyrinthine halls and rooms within a major gallery. Darkness and enlightenment are loaded terms. Experiencing art is not the same as knowing about it, and this anti-Exhibition challenges the orthodoxy of presenting art as a series of connected events. Rather than showing Picasso after Cézanne, reflecting the history and order of tradition, Høyersten sets out to explore other possible contemporary meanings, other relationships between treasured art objects in ways that unsettle "order and harmony" and make the viewer look afresh at art. The gallery is an inside-out space that deliberately disorients and redirects the viewer. A room of thick fog lit in blue-purple unbalances the visitor and confuses perceptual depth cues. Emerging from the fog-room, she finds herself in the heart of the show, a huge *Wunderkammer*. She is amid ancient and modern art objects, juxtaposed on standard backroom museum shelving from high to low all right around her. Especially noticeable are the small Andy Warhol works of Marilyn Monroe sitting on shelves all around the room, no longer a familiar set or a series but broken apart. The assortment of little Marilyn Monroes has been juxtaposed with ancient Egyptian or Danish tomb finds, or with a practical objet-d'art from another century. Paintings are arrayed on shelves alongside ceramic, stone, bone, horn, and bronze objects. Time is in turmoil here, but each object is lit for appreciation.

"To present art without structure," is, according to Høyersten, to "fly in a cloud without instruments." The *Wunderkammer* in ARoS invites the individual to "pick and shop and zap," to view work without signposts, to see art straight *Out of the Darkness*. Høyersten is a curator, not a creator of fine art: his skill is in the mediation. His chosen objects are tools to think with, and his "pick and shop and zap" approach engages with distracted viewers with the poor attention spans of our digital age. Where visitors are used to learning from screens and spend little time with any one item, they need fresh ways to find mindfulness in the sacred space that ARoS offers. The cabinet here enabled a set of physical objects to be read as hypertext; the fog that preceded the room opened up the mind to the immediacy of delving without prescription, to follow one's unconscious in responding to the objects arrayed there. Despite the fragmentation, the total experience of the *Wunderkammer* is larger than the sum of the parts. Høyersten demands the viewer's own disoriented body to provide the cabinet's organizing principle. She herself becomes the source of wonder, providing the curiosity that makes the cabinet.

WRONG WAY TIME

The Australian artist Fiona Hall uses an Aboriginal English term "Wrong Way" to tackle the question of Time in her stunning retrospective/prospective exhibition. *Wrong Way Time* (2012–15), curated by Linda Michael, was first shown at the Australian Pavilion in Venice, part of the 56th International Art Exhibition la Biennale di Venezia May 9–November 22, 2015, then at the National Gallery of Australia, Canberra, April 22–July 10, 2016. The Aboriginal reference comes from the exhibition within the exhibition, *Kuku Irititja* (*Animals from Another Time*), a 2014 collaboration undertaken by Hall in partnership with the Tjanpi Desert Weavers of the NPY country in central Australia: forty sculptural works woven with *tjanpi* wild grasses, camouflage military garments, feathers, cans, and burnt volumes of the *British Museum's General Catalogue of Printed Books 1956–1965*. These dates are significant, the years of British nuclear testing in NPY lands. The exhibition celebrates the animals of a happier time, many of which are now extinct. They come from a time before they all got sick: "We were not sick before the nuclear fallout landed on us" (Hall 2015, 54). Living "wrong way" in the new times since the nuclear moment is living in the Anthropocene.

The Hall exhibition is a suite of different environmental and justice projects displayed in cabinets of wonder, surrounded by clocks—grandfather clocks, cuckoo clocks, mantle clocks, banjo clocks, covered in tally marks and skeletons. These the unforgiving instruments of time line the walls of the room, chiming, ticking and donging constantly and unharmoniously, *Counting for Nothing* (2014), as one grandfather clock declares.[2] The oppressive timepieces create the soundscape for the cabinets of wonder at the heart of the display, cabinets of curious objects, created for different projects spanning the quarter century from 1990–2015. Hall defies description as an artist: she is a sculptor, a fine artist, and a philosopher who works with sardine tins, Coke cans, and banknotes, with 3D installations and with society's excesses. She makes tiny and enormous forms, all exquisite. *Tender* (2003–6), on loan from the Queensland Art Gallery, is eighty-six works of woven US banknotes, precision models of birds' nests. Cabinet D (both sides) is devoted to *Tender*. Art historian David Hanson describes the vacuum in which this works. It is not so much artifice as *emptiness*:

> Not only are there no birds—no parents, no chicks, no eggs, even. There are no trees—no protection, no photosynthesis, no oxygen respiration. Just the faintest echo of faded chlorophyll in the lichen or lawn-clipping green of the paper currency itself. There is, in fact, no environment, other than the abstract museum space of the vitrine. (Hanson 2015, 46)

Beyond Clutter

Cabinets of Curiosity in Worm's day lacked the timing and historical precision that informs Fiona Hall's art. Høyersten actively disrupts the art-history sequence in order to force a rethinking of the historical systems to re-organize his art museum. Worm and others in the pre-Enlightenment *Wunderkammer* era collected material objects from nature and "other" (ethnographic) cultures, and arrayed them by morphology or geography. More was better. In our present era of overconsumption, we no longer experience the awe of too much. It is commonplace, at least in the western world. The aim of our new cabinets is to recreate the curious, to extract meaning from excess, to create mindfulness with silence or soundscapes, with fog, darkness and hidden lighting deep in boxes.

Clutter is no longer art. Rather there is a demand to look again, to rethink what we have already, and perhaps what we have lost. Høyersten's ARoS challenges the orthodoxy of art curation: *Out of Darkness* is his manual for reading the whole museum, from the oversize figure of a frightened boy, hiding (in full view) in the basement to the penthouse gallery of rainbow windows on the city of Aarhus. The whole museum is a cabinet of sorts, a place apart from the city that comments on it, and offers respite from it. Hall has created masterly new art from recycled, troubled matter. She has taken the discarded and remade works of beautiful art. Anthropocene cabinets grapple with life in a "no analogue" world beyond the Holocene (Crutzen and Steffen 2003, 253). Her art, like life in the Anthropocene, juggles multiple intellectual systems and the interactions between them. The Anthropocene is not about any one sort of global change, but how climate change, changes in biosphere integrity, atmospheric pollution, ocean acidification, shifting economic systems, and social injustice all work together and change each other through feedback loops. Hall also debates justice, environment, extinction, and nuclear fallout using sculpture, painting, weaving, camera obscura, and trompe l'oeil technologies. The Anthropocene is a geological concept invented by chemists and ecologists, but imagining it, living in it, and adapting to its unknown unknowns demands a new humanism, too. The Great Acceleration of changes undoubtedly adds trauma and stress to our present era: the digital revolution, global banking and financial systems, and food production are all changing fast and interdependently, putting pressure on individual living and global communities.

The story of climate change is most commonly told through sciences that record statistical rises in temperature and strange weather, through the acidification of oceans and CO_2 levels in the atmosphere. There are stratigraphers looking for traces in the rocks that might constitute "markers" of the Anthropocene in a possible future world where there are no longer humans at all. How might a device, a *Wunderkammer*, explore *adaptive* responses to accelerating and interacting changes? The Anthropocene is much more than climate change, although this is a big driver. It has the potential to be so overwhelming that it threatens not just our future, but also our ways of aspiring to quality of life in the future. While climate scientist Mike Hulme (2011, 245) urges that we must not reduce the future to climate, anthropologist Arjun Appadurai (2013, 286) reminds us that the future is cultural. One of the greatest inequalities of the world is between those that can aspire to a future and those that cannot. Humanity cannot expect to revert to Holocene conditions, the epoch in which most of the world's most prominent present civilizations emerged and flourished. With different environmental and climatic conditions, civilizations will be different too, if they (and we) survive. While there are humans, conscious and making sense of Earth, human stories remain important.

Cabinets of Connectivities

Objects can be very useful in bringing global ideas back to a human scale and making them personal, but there are still choices to be made. The western world is now overloaded with meaningless material goods. An object was an identifiable and precious thing in the first era of the *Wunderkammer* where an average middle European household had about thirty objects. Each one was treasured and noticed (Trentmann 2016). Each object had its own story: "this came from that tree," or a grandparent, or a moment of necessity. In the twenty-first century the number of household objects has risen to 12,000. No one can recall or tell stories for them all, so they are just "stuff." In our era where people live in bigger houses, encumbered with more and more stuff, can the older practice of looking carefully at "curious objects" surprise us afresh and help with the anxiety of living in times of rapid environmental and social change?

When ordinary people have too many objects, how can each become special again? How can they evoke care, mindfulness—even fresh wonder and curiosity? Is it a mark of apocalyptic times that photographers delight in "ruin porn," photographing the desolate sites of post-industrial Detroit? Cabinets of Curiosities challenge the routine screening out of "too much information" that is the malaise of our times. From an early age we sort our world into what we notice and what we don't. As we get older we become so proficient at not-noticing, we race through life without seeing the world at all. It may take a small child to remind an adult to delight in raindrops on a leaf. Scientifically, we have a taxonomy of knowledge—not just the classifications of plants and animals in the Linnaean system, but also the ways we do that classifying. Only some morphological characteristics define groups, most are irrelevant. During the Enlightenment, the museum cabinet displayed traditional orders of knowledge: an array of butterflies or a cabinet of African animals (ordered by habitat) or Ungulates (ordered by morphology). DNA-based cladistics may suggest different family stories and suggest new orders. Cryogenic collections and deep-frozen cabinets have another order again.[3] Depending on the expertise of the observer, the order of nature is different. In our postsystematic era, rearranging objects in cabinets is more art than science.

In an ecological world that now eschews the idea of balance in nature, nature is full of "Discordant Harmonies," according to ecologist Daniel Botkin (1990). Chaos theories inform our understandings of nature, and Cabinets of the Future grapple with this. Some of the most exciting new cabinets are crossovers between arts and science, between curatorial creativity and systematic rigor. The Deutsches Museum's *Welcome to the Anthropocene* exhibition (2014–16) is itself built in modules. These—Cabinets of Curiosity in another guise—draw on the work of artist Yesenia Thibault-Picazo to explore the idea of "future geology."[4] Thibault-Picazo produced her *Anthropocene Specimen Cabinet*, a cabinet of future fossils, in collaboration with geologist Jan Zalasiewicz, chair of the Anthropocene Working Group for the Stratigraphy Commission of the Geological Society of London. Through her "do-it-yourself Geomimicry," she built specimens of the most anthropogenic materials and accelerated their fossilization through ovens and heat presses, creating metallic rocks, for example, from discarded mobile phones, and PPC Mortar of the Pacific Plastic Crust—that gyre of rubbish in the Western Pacific Ocean. Thibault-Picazo's work unsettles and revitalizes the emotions in engaging with life in the Anthropocene. She derives her ideas about the future from history and was inspired by Georgius Agricola's 1556 Treatise *De Re Metallica* in works like her *Future Mining Tools* series pick/fan/rake/ruler. Rather than providing a dull and technical commentary on the "end of the Holocene," she suggests a beautiful geological future. In the shadows of her objects, there is hope in times of a new geological order.

Curating Connections—Performing the Everyday

The case of the ARoS *Wunderkammer* reminds us that museums are places of performance as well as passive object displays. The staging of the event (or the visit) to enable the revelation of an artefact is the work of the curator. To understand the curatorial imagination demands getting right *inside* the cabinet: not just looking through the cabinet window. From inside a *Wunderkammer*, a viewer can, paradoxically, think "outside the box." A cabinet is a microcosm, and a tool to scale up and down time and space. In *Future Remains* we have a cabinet, too, replete with "fragmentary histories," objects that evoke our Anthropocene moment, created originally through performance. Through an event akin to a poetry slam in Madison, Wisconsin, the objects of this book first encountered their "cabinet companions." Eight months later they reassembled in another cabinet in Munich, Germany—where they became both a performative event in July 2015, and a Cabinet of Curiosities complementing the main display in the Deutsches Museum gallery, *Welcome to the Anthropocene* (Robin and Muir 2015). The objects became symbols of the participatory and inclusive philosophy advocated for the Anthropocene gallery by its chief curator Nina Möllers. Möllers aimed for a "playful" gallery that enabled hope and personal response (Möllers 2015a, 108–12; Robin et al. 2014).

The Deutsches Museum's *Welcome to the Anthropocene* exhibition is a gallery of conceptual cabinets with objects. The visitor is enticed to explore a big idea one cabinet at a time, part of a deliberate strategy to move away from an "authoritative" exhibition to an inclusive and active one. The Wall of Anthropocentric Objects is loaded with objects from the museum including a classic steam-engine, the object that Paul Crutzen identified as the starting point for accelerating carbon in the atmosphere (2002, 23). The Wall can be viewed from front (past) or back (future) or *within* the arched doorway that takes one into a space of jigsaw island modules, each exploring a key theme. Each island is a cabinet of individual curatorial visions, with distinctive conceptual styles, despite a common architecture of white and blue. *Food* is set around an oversized table, and the visitor has to decide what she will eat, guided by many of the big debates in contemporary German politics. *Global Plant Movements* is more international and historical; undoubtedly alien invasive species are something keenly felt and more highly politicized in neo-Europes. The six "island" cabinets (or modules) are complementary, literally: the aerial view of them is a single continent, broken apart as pieces of a jigsaw, allowing each visitor to reassemble the whole in their own way. My favorite element of the cabinets is the discovery drawers in the outside of each island, a "behind the scenes" dimension, and a physical acknowledgment that we approach complex concepts both from within and without.

This gallery entertained and inspired, rather than provided information about a no analogue future to a passive viewer. Viewers had to actively make their experience of the Anthropocene future, and reflect on how and what one chose to have there. The design emphasized the qualitatively *possible*, above gloomy predictions of the quantitatively *probable*: it spoke of a future beyond numbers (Robin 2015).

A Cabinet of Curiosities challenges the order of things through juxtaposition. Whose systems are ordered through the *Wunderkammer*? How might the cabinet concept foster a greater inclusiveness for different, even incommensurate systems? Can the readings of the original curator be different from those of the viewer? What about the ordinary—what place does it have in exploring the extraordinary?

The role of the curator in a history museum is to explore these questions; to display found objects, rather than art works. The role of the viewer is to read the objects for stories—original, and perhaps also their own. Objects can catch viewers in unexpected ways, triggering memories, revelations, understandings beyond what the curator had envisaged. That is the magic of museums. There is curatorial practice, and there is the practice of the visit itself: the professional and the amateur work together, conjuring experience with objects. The moment of slow meditation that the visit enables is akin to a reflective moment in a secular cathedral; intellectual and emotional understandings come together. This book is about a collection of found objects, discovered by different curators and performed in the ritual of a "slam" event. The objects finally came to rest in a Cabinet of Curiosity, an adjunct to an artful exhibition of the Anthropocene half a world away from their first performance. Now these stories continue around the world, in this book of *Future Remains*.

Exploring the Ordinary at the End of the World

I return to Denmark to reflect on the everyday at the end of the world. Denmark is a low-lying country that finally subsides into not one, but two seas in its far north. At the end of a long sand spit, the Baltic Sea collides with the North Sea, blue on grey, with whale-like spurts of white surf along a near straight line. Light is important—even magical in such northern climes where long dark winters inspire their own noir-thrillers. Skagen, the town near the last spit of sand, is famous as the home of the artists of the Golden Era of Danish painting.

Ends of the earth have fascinated me since I was a child reading C.S. Lewis's *Chronicles of Narnia*: I have always wanted to go to Cair Paravel (Lewis 1956). In my imagination it was at the vanishing point where sea and sky merge, on the horizon of an Australian beach in summer. In our Anthropocene age, big global systems of sky and sea—the warming atmosphere and the acidifying oceans—are also associated with the end of the world, at least the one we have known in the Holocene. As I wandered along the lonely Danish North Sea strand toward the end of land in the late summer of 2015, I reflected on the power of abstract, unimaginable ideas to influence economics to stop "discounting the future," and to give space for both the human and the other-than-human lives in the emerging novel ecosystems of our times.

The dance between the human and the natural is different when the curator is the wind, or circumstance, not an act of human artistry. The creative work passes entirely to the viewer of the cabinets here. In this beach windscape I encountered a stranded boat, no longer sea-faring, marked out from the sand-drift with an inexplicable line of dead Christmas trees. In its belly was a sandy floor and brackish yellow-green water, sourced from high tides and heavy rains. Kelp and small swimmers already inhabited this unlikely Ark of the Future. This novel ecosystem has been curated by circumstance, built with human detritus by living creatures of all sorts.

From the prow of the boat I lifted my eye to the sea shore. Gentle wavelets were lapping an ephemeral island of the North Sea. It is an experimental place like the islets of the Florida Keys where Robert MacArthur and Edward O. Wilson developed their theory of island biogeography in the 1960s, a theory that shaped the way national parks were managed for much of the rest of the century. I spotted the simplest possible island. It had no vegetation and a life span of just the few hours between the tides. Yet already its shore was inhabited by the first-and-last of life, a glorious red jellyfish.

Finally, I looked along the sand where my footprints stretched back several kilometers. Here the dried-out entrails of kelp created improvised containers—perhaps even cabinets—for flotsam and jetsam collected by the tidal fringes. The kelp vessels were surprisingly pretty decorated with colorful scraps of plastic rope—red, blue, yellow—no longer needed by sailors and now unravelling into the white sand and grey stones of the sea shore. All the way along this beach little kelp cabinets paid tribute to the interconnectedness of wild life and twenty-first-century human technologies. Every time I spotted a scrap of rope I saw also the feathers of sea birds, caught by rope; now their final resting places were decorated with kelp, rope, sand drift, and stones. Rope and birds were both stranded from the ocean: they died entwined together. There were no feathers without rope, and no rope scraps without feathers.

The Environmental Humanities: Museums, Creative Scholars, and Communities

Objects are powerful tools, engaging emotional responses including grief and uncertainty and other negative feelings about change. There is a fruitful public intellectual space between universities and museums where all sorts of practitioners of the environmental humanities can explore life at the end of the Holocene.

A question for museums is what *should* one collect for the future, if one's place is changing beyond recognition? Museums have a particular role in helping communities engage positively with their cultural futures, not merely practical adaptations to biophysical change. But what is a community in this age of change? Some communities are composed of globally distributed people, exposed to a glorious variety of other ways of life, yet paradoxically narrowed by a sameness of purpose. Others value their local place, but find it diffuse, dispersed, and dis-*placed*.

Losing a place, emotionally if not actually, creates sadness, a sort of homesickness, that environmental psychologist Glenn Albrecht (2012) has described as "solastalgia": you may be living on the same land, but it is no longer the *place* of your forebears. The idea of losing place has inspired and stimulated many artists. Masao Okabe and Chihiro Monato created an "archive of the future," using archive boxes to store documents and artworks to "keep in memory" the Ujina railway station in Hiroshima.[5] Ujina is a place that has been doubly lost by atomic warfare and by the twenty-first-century redevelopment of the city: it was bombed in 1945 and dismantled in 2002. *Hiroshima in Tasmania*, developed originally for the Japan Pavilion of the 52nd Venice Biennale, is an artistic archive that was built on the remaining stones from the railway platform. It has now found a permanent home at MONA, the Museum of Old and New Art, in Hobart, Tasmania. Along with the archive boxes, there is a performative task: an invitation to make new frottages by rubbing the stones "for the future," to reactivate the global memory of this iconic lost place. If as, is likely, the stratigraphers decide that the Anthropocene should be dated from the traces of nuclear explosions in the rocks following the nuclear explosions of the 1940s, this archive of the future will acquire another, Anthropocene, significance.

People feel sadness, dislocation, thrill, and challenge as they live in Anthropocene times. The Great Acceleration of Time is disorienting, too—not just loss of place, but loss of time to be mindful is the experience of the racing, digital age. Exploring "curiosities" and imagining the futures of great-great-grandchildren can steady such heady emotions. An object stays still. It can be held in a present reality in the chimera of experience across time. In exploring the moral and personal dimensions of a geological epoch predicated on the concept of humans as drivers of biophysical processes on a global scale, the environmental humanities traditions of art and observation, of history and poetry, work together. This essay suggests another craft of the environmental humanities, curation, which is making a revival in times of strange change through Cabinets of Curiosity. Engaging with curious objects is a task of art and science. It is both natural and unnatural history, creating a possibility of an Anthropocene beyond science. Engaging unconscious as well as conscious perceptions of change, and allowing "other than human" leadership in the curating project, is a multimedia, multidisciplinary task. Curation is a skill for natural history museums, social history museums, for high and low art traditions, for justice and for the wider world. The curious juxtapositions and connectivities of the *Wunderkammer* are suggestive of ways of thinking differently for our strange and uncertain times.

NOTES

I would like to acknowledge many stimulating conversations with curators including Jennifer Newell, Kirsten Wehner, Christine Hansen, Nina Möllers, Luke Keogh and Helmuth Trischler, also with my colleagues in projects about the history of expertise for the future, Sverker Sörlin, Paul Warde, Frank Trentmann and Rebecca Wright, and all the KTH Environmental Humanities Lab, Stockholm. Special thanks to artists whose works enable us to aspire to a rich cultural future, and Anthropocene scientists including Will Steffen and Jan Zalasiewicz, whose humanism informs the way they do science. I am particularly grateful to the editors of this volume and readers of earlier drafts of this chapter for stimulating comments and suggestions.

1   Erlend Høyersten (a Norwegian) was appointed director of ARoS in 2014, and this exhibition was a statement of his style. Wording here comes from the signboard in the exhibition, which has now closed. Some details remain here: http://en.aros.dk/exhibitions_/2014/out-of-the-darkness/ (accessed July 16, 2016).
2   Detail from Hall 2015, image facing p. 56.
3   Kowal and Radin 2015, 63–80; Also the panel "Lost Species: Extinction in Museums," chaired by Kirsten Wehner, with speakers Joshua Drew, Joanna Radin, Kathryn Medlock, Nancy Simmons, and George Amato at Collecting the Future, Symposium at American Museum of Natural History, October 2013, http://www.amnh.org/our-research/anthropology/news-events/collecting-the-future/.
4   Möllers 2015b, 167; Y. Thibault-Picazo, "Craft in the Anthropocene: Fossils from the Future," in Möllers, Schwägerl, and H. Trischler 2015, 114–19. See also Y. Thibault-Picazo, "Future Geology," https://vimeo.com/68612912.
5   Text is available here: http://www.writingtheearth.com/2011/10/is-there-future-for-our-past.html.

BIBLIOGRAPHY

Albrecht, G. 2012. "The Age of Solastalgia." *The Conversation*, August 7. https://theconversation.com/the-age-of-solastalgia-8337.
Appadurai, A. 2013. *The Future as a Cultural Fact: Essays on the Global Condition*. London: Verso.
Botkin, D. 1990. *Discordant Harmonies: A New Ecology for the Twenty-First Century*. New York: Oxford University Press.

Crutzen, P. J. 2002. "Geology of Mankind." *Nature* 415 (6867): 23.

Crutzen, P. J., and W. Steffen. 2003. "How Long Have We Been in the Anthropocene Era? An Editorial Comment." *Climatic Change* 61 (2003): 253.

Griffiths, T. 1996. *Hunters and Collectors: The Antiquarian Imagination in Australia.* Cambridge: Cambridge University Press.

Hall, F. 2015. *Wrong Way Time.* Edited by L. Michael. Sydney: Piper Press/Australian Council for the Arts.

Hanson, D. 2015. "The Folding Stuff." In F. Hall, *Wrong Way Time,* ed. L. Michael, 39–47. Sydney: Piper Press/Australian Council for the Arts.

Hulme, M. 2011. "Reducing the Future to Climate: A Story of Climate Determinism and Reductionism." *Osiris* 26:245–66.

Impey, O., and A. MacGregor, eds. 1985. *The Origins of Museums: The Cabinet of Curiosities in Sixteenth- and Seventeenth-Century Europe.* Oxford: Clarendon Press.

Kowal, E., and J. Radin. 2015. "Indigenous Biospecimen Collections and the Cryopolitics of Frozen Life." *Journal of Sociology* 51 (1): 63–80.

Lewis, C. S. 1956. *The Last Battle.* London: Bodley Head.

Möllers, N. 2015a. "Museums and the Anthropocene: Reconfiguring Time, Space and Human Experience." In *Welcome to the Anthropocene: The Earth in Our Hands,* ed. N. Möllers, C. Schwägerl, and H. Trischler, 108–12. Munich: Deutsches Museum and Rachel Carson Center.

Möllers, N. 2015b. "The Nature of the Future." In *Welcome to the Anthropocene: The Earth in Our Hands,* ed. N. Möllers, C. Schwägerl, and H. Trischler. Munich: Deutsches Museum and Rachel Carson Center.

Möllers, N., C. Schwägerl, and H. Trischler, eds. 2015. *Welcome to the Anthropocene: The Earth in Our Hands.* Munich: Deutsches Museum and Rachel Carson Center.

Robin, L. 2015. "A Future Beyond Numbers." In *Welcome to the Anthropocene: The Earth in Our Hands,* ed. N. Möllers, C. Schwägerl, and H. Trischler, 19–25. Munich: Deutsches Museum and Rachel Carson Center.

Robin, L., and C. Muir. 2015. "Slamming the Anthropocene: Performing Climate Change in Museums." *reCollections* 10 (1). http://recollections.nma.gov.au/issues/volume_10 _number_1/papers/slamming_the_anthropocene.

Robin, L., D. Avango, L. Keogh, N. Möllers, B. Scherer, and H. Trischler. 2014. "Three Galleries of the Anthropocene." *Anthropocene Review* 1 (3): 207–24. doi:10.1177/2053019614550533.

Trentmann, F. 2016. *Empire of Things: How We Became a World of Consumers, from the Fifteenth Century to the Twenty-First.* London: Allen Lane.

Contributors

MARCO ARMIERO is the director of the Environmental Humanities Laboratory at the KTH Royal Institute of Technology, Stockholm, where he is also an associate professor of environmental history. He is the author of *A Rugged Nation: Mountains and the Making of Modern Italy* (2011) and is associate editor of the journal *Environmental Humanities*. His articles have been published in *Left History*, *Radical History Review*, *Environment and History*, *Modern Italy*, and *Capitalism Nature Socialism*, where he also serves as a senior editor. His edited volumes include *Nature and History in Modern Italy* (with Marcus Hall) and *The History of Environmentalism* (with Lise Sedrez). His new edited volume, *An Environmental History of Modern Migrations* (with Richard Tucker), was published by Routledge in 2017.

TRISHA CARROLL was born in Grenfell, New South Wales, and is a Wiradjuri elder or "auntie." She has taught the Aboriginal culture and literacy program for high schools in Cowra and has also been an Aboriginal Land Council Representative for Cowra region. She is an artist and award-winning country and western musician. She has exhibited nationally in regional and commercial galleries. She has shown at Roslyn Oxley9 Gallery, Sydney and Christine Abrahams Gallery, Melbourne and internationally at the Deutsches Museum, Munich, in collaboration with fellow artist Mandy Martin. Carroll has presented at Indigenous and environmental conferences, workshops, and forums in collaboration with Mandy Martin.

ROBERT S. EMMETT is the author of *Cultivating Environmental Justice: A Literary History of US Garden Writing* (University of Massachusetts Press, 2016) and with David Nye, *Environmental Humanities: A Critical Introduction* (MIT Press, 2017). His work has also appeared in *Interdisciplinary Studies in Literature and Environment* and *Environmental Humanities*. From 2013 to 2015 he served as director of programs at the Rachel Carson Center for Environment and Society in Munich, Germany. He is currently visiting assistant professor of environmental studies at Roanoke College, Virginia.

JARED FARMER teaches history at Stony Brook University. He is the author of *On Zion's Mount: Mormons, Indians, and the American Landscape* (Harvard University Press, 2008), winner of the Francis Parkman Prize, and *Trees in Paradise: A California History* (Norton, 2013), winner of the Ray Allen Billington Prize. In 2014 Farmer received the Hiett Prize in the Humanities. Visit his website at http://jaredfarmer.net.

NILS HANWAHR is a doctoral student at the Rachel Carson Center for Environment and Society, run by the Ludwig-Maximilians-Universität Munich and the Deutsches Museum München. He holds master's degrees from the University of Oxford and Imperial College London. Prior to his doctoral studies he worked in science policy and as a product manager for inflight satellite connectivity.

RACHEL HARKNESS is a lecturer in critical and contextual studies in the School of Design, Edinburgh College of Art, University of Edinburgh. Her research explores architecture and design as a peopled process, and considers how people make manifest their (eco-)designs for living. She writes on art, space-time, value, materials, the senses, and the environment. Her recent work considers vibrant materials in the built environment, and the stories, entanglements, and skills that a focus upon them brings to the fore. It does so, in part, through playful experiment and participation in artistic practices of making.

NICOLE HELLER is a conservation ecologist and expert in the field of climate change adaptation for biodiversity protection and urban resilience. She has extensive experience practicing informal science education through collaborations with artists, media and citizen scientists. She has served as an assistant professor at Duke University, and at Franklin and Marshall College, and also as a researcher and science director with numerous environmental nonprofits. Nicole's research has been published in both academic and popular presses. She has a PhD from Stanford University, a BA from Princeton University, and was a postdoctoral fellow at University of California, Santa Cruz.

ELIZABETH HENNESSY is an assistant professor of history and environmental studies at the University of Wisconsin–Madison, where she teaches on animal histories and the history of the Anthropocene. Trained as a geographer, she works at the intersection of environmental history, political ecology, science and technology studies, and animal studies. Her first book tells the story of the giant tortoises of the Galápagos Islands to examine the contemporary politics of evolutionary understandings of nature.

JUDIT HERSKO is an installation artist who works in the intersection of art and science and collaborates with scientists on visualizing climate change science through art. Her work has been featured internationally, including in the 1997 Venice Biennale where she represented her native Hungary. In 2008 she received the National Science Foundation Antarctic Artists and Writers Grant and spent six weeks in Antarctica. Her performance lectures in the series Pages from the Book of the Unknown Explorer build on this experience and investigate effects of climate change, as well as the history of polar exploration and science.

GARY KROLL received his PhD from the University of Oklahoma. He is professor of history at SUNY Plattsburgh, where he specializes in modern US environmental history and the history of science. Gary is the author of *America's Ocean Wilderness: A Cultural History of Twentieth Century Exploration*, coauthor of *Exploration and Science*; and coeditor of *World in Motion: The Globalization and the Environment Reader*. His current project is a cultural history of the collision between fossil-fuel-based transportation and American wildlife.

MICHELLE MART is author of *Pesticides, A Love Story: America's Enduring Embrace of Dangerous Chemicals*, a cultural history of pesticide use in the United States from World War II to the present. She is working on a study about food, culture, and the environment. Previously, she wrote about culture and foreign policy in *Eye on Israel: How America Came to View Israel as an Ally*. In 2012 and 2014 she was a Carson Fellow at the Rachel Carson Center for Environment and Society in Munich and is currently associate professor of history at Penn State University, Berks Campus.

MANDY MARTIN is an artist who has held numerous exhibitions in Australia and internationally. Her works are in many public and private collections including the National Gallery of Australia, the Art Gallery of New South Wales, and other state collections and regional galleries. In the United States she is represented in the Guggenheim Museum New York, the Los Angeles Museum of Contemporary Art, the Nevada Museum of Art, Reno, and many private collections. She lives in the Central West of New South Wales, Australia, and is currently an adjunct professor at the Fenner School of Environment and Society, Australian National University. http://www.mandy-martin.com

JOSEPH MASCO is a professor of anthropology and science studies at the University of Chicago. He is the author of *The Nuclear Borderlands: The Manhattan Project in Post Cold War New Mexico* (Princeton University Press, 2006) and *The Theater of Operations: National Security Affect from the Cold War to the War On Terror* (Duke University Press, 2014). He is currently working on several projects exploring the intersection of planetary scale insecurities, ecological relations, and critical theory.

TOMAS MATZA is an assistant professor in the Department of Anthropology at the University of Pittsburgh. His work focuses broadly on the political economy of crises and their aftermath, with particular emphasis on mental health and environment. His first book is entitled *Shock Therapy: The Ethics and Biopolitics of Psychotherapy in Post-Soviet Russia* (Duke University Press, forthcoming). His work has been featured in *Cultural Anthropology* and *American Ethnologist*. He has also coauthored an article, "Putting Foucault to Use," with Colin Koopman, published in *Critical Inquiry*, and coedited a special issue of *Social Text*, "Politically Unwilling," with Kevin Lewis O'Neill.

DAEGAN MILLER earned his PhD in history from Cornell University and was an AW Mellon Postdoctoral Fellow in the Humanities at the University of Wisconsin-Madison. He is now a writer, and his work has appeared in a wide array of venues, from literary magazine to academic journals, including the *American Historical Review*, *3:AM Magazine*, *Environmental Humanities*, and *Stone Canoe*. His first book will soon be published by the University of Chicago Press. Find out more about Daegan at http://daeganmiller.com.

GREGG MITMAN is the Vilas Research and William Coleman Professor of History of Science, Medical History, and Environmental Studies at the University of

Wisconsin–Madison. His work spans the history of science, medicine, and the environment in the United States and the world, and reflects a commitment to environmental and social justice. His most recent books include *Documenting the World: Film, Photography, and the Scientific Record* (coedited with Kelley Wilder) and *Breathing Space: How Allergies Shape our Lives and Landscapes*. Together with Sarita Siegel, he directed and produced *The Land Beneath Our Feet*, a documentary on history, memory, and land rights in Liberia, available through Passion River Films.

CAMERON MUIR is a research fellow at the National Museum of Australia and was previously a fellow at the Rachel Carson Center for Environment and Society, Munich. He is the author of *The Broken Promise of Agricultural Progress: An Environmental History* (Routledge, 2014).

ROB NIXON holds the Currie C. and Thomas A. Barron Family Professorship in Humanities and the Environment at Princeton University. He is the author of four books, most recently *Slow Violence and the Environmentalism of the Poor*, which won numerous awards, including the 2012 Sprout Prize from the International Studies Association for the best book in environmental studies. Nixon writes frequently for the *New York Times*. His writing has also appeared in the *New Yorker*, the *Atlantic*, the *Guardian*, the *Nation*, *Chronicle of Higher Education*, *London Review of Books*, and *Critical Inquiry*.

LAURA PULIDO is a professor in the departments of Ethnic Studies and Geography at the University of Oregon where she focuses on environmental justice, race, and critical human geography. She is the author of numerous books including *Environmentalism and Economic Justice: Two Chicano Struggles in the Southwest, Black, Brown, Yellow and Left: Radical Activism in Los Angeles*, and *A Peoples Guide to Los Angeles* (with Laura Barraclough and Wendy Cheng), a radical tour guide documenting sites of racial, class, gender, and environmental struggle in the landscape and history of Los Angeles.

LIBBY ROBIN is a historian at the Fenner School of Environment and Society, Australian National University. She is also affiliated professor at KTH Stockholm and the National Museum of Australia. She contributed to the first "Anthropocene Slam" in Madison, Wisconsin, in November 2014 and moderated a CLIMARTE slam in Melbourne, May 2015. She is coeditor (with Jennifer Newell and Kirsten Wehner) of *Curating the Future: Museums, Communities and Climate*

Change (Routledge 2017), and has published widely on the Anthropocene. She was elected Fellow of the Australian Academy of Humanities in 2013.

CRISTIÁN SIMONETTI is assistant professor at the Programa de Antropología, Instituto de Sociología, Pontificia Universidad Católica de Chile and an Honorary Research Fellow at the Department of Anthropology, University of Aberdeen. He has conducted fieldwork with land and underwater archaeologists in Chile and Scotland, and with glaciologists in Greenland. His research concentrates on how scientists studying the past understand time, the topic of a monograph he is currently writing for Routledge, entitled *Sentient Conceptualisations*. Recently he has been involved in collaborative explorations, across the sciences, arts, and humanities, on the properties of inorganic materials such as ice and concrete.

SVERKER SÖRLIN is professor in the Division of History of Science, Technology and Environment at KTH Royal Institute of Technology, Stockholm, and cofounder of the KTH Environmental Humanities Laboratory. His current work is on the science politics of climate change and the Anthropocene. He has published narrative nonfiction, essays, journalism, and scripts for television and film. He is also an expert and government adviser on research policy and the future of the humanities. He is coeditor of *The Future of Nature* (Yale University Press, 2013) and coauthor with Libby Robin and Paul Warde of *The Environment: A History* (forthcoming).

JULIANNE LUTZ WARREN has a PhD in wildlife ecology. She is author of *Aldo Leopold's Odyssey, Tenth Anniversary Edition*, which unfolds this eminent American conservationist's land health vision of humans dwelling generatively within the world-of-life. Julianne's other scholarly and creative works focus on possibilities for living that out. While teaching at New York University, Julianne won a Martin Luther King Jr. Faculty Research Award for her still-ongoing work with students in the climate justice movement. She is a contributor to the Land Institute's Ecospheric Studies working group and serves as a Fellow of the Center for Humans and Nature.

BETHANY WIGGIN is the founding director of the Penn Program in Environmental Humanities, Topic Director for Translation at the Penn Humanities Forum, and Graduate Chair in German. She has published on the rise of the novel and reading as entertainment in Europe; fears of fashion and commodity culture, including slavery; and multilingual and world literature. She directs the WetLand

Project, a public collaboration to witness a river's pre-and postcolonial, pre- and postindustrial past and present—and to imagine its futures. She is writing *Utopia Found, Lost, and Re-Imagined in Penn's Woods.*

JUDITH WINTER is a curator, writer and senior lecturer at Manchester School of Art, Manchester Metropolitan University. She has worked both within and beyond the gallery context, most notably as inaugural curator for the Middles-brough Institute of Modern Art (MIMA), UK and as Head of Arts for Dundee Contemporary Arts (DCA), Scotland. She has facilitated solo exhibitions by many contemporary artists, including Thomas Hirschhorn (Switzerland), Martin Boyce (Scotland + Venice Biennale); Johanna Billing (Sweden), Matthew Buckingham (United States), and Mafred Pernice (Germany).

JOSH WODAK is a researcher, artist, and design educator in the Faculty of Art and Design, University of New South Wales. His work critically engages with cultural and ethical entanglements between environmental engineering and conservation biology as means to mitigate species extinction and biodiversity loss in the Anthropocene. He holds a BA (Honours) in anthropology (Sydney University, 2002), a PhD in interdisciplinary cross-cultural research (Australian National University, 2011) and is currently a chief investigator on the Australian Research Council Discovery Project *Understanding Australia in The Age of Humans: Localising the Anthropocene* and a member of the Andrew Mellon Australia-Pacific Observatory in Environmental Humanities, Sydney Environment Institute, University of Sydney.